Numerical and experimental investigations of distribution of gaseous emissions with the air flow in the indoor environment

Von der Fakultät Energie-, Verfahrens- und Biotechnik

der Universität Stuttgart zur Erlangung der Würde

eines Doktor-Ingenieurs (Dr.-Ing.)

genehmigte Abhandlung

vorgelegt von

Umer Afzal

geboren in Lahore, Pakistan

Hauptberichter: Prof. Dr.-Ing. habil. Clemens Merten

Mitberichter: Prof. Dr.-Ing. Philip Leistner

Tag der mündlichen Prüfung: 14.06.2017

Institut für Chemische Verfahrenstechnik

der Universität Stuttgart

2017

Bibliographic information published by the Deutsche Nationalbibliothek

The Deutsche Nationalbibliothek lists this publication in the Deutsche
Nationalbibliografie; detailed bibliographic data are available
on the Internet at http://dnb.d-nb.de .

Printed and/or published with the support of
the Higher Education Commission (HEC) Pakistan
and the German Academic Exchange Service (DAAD)

ISBN 978-3-8325-4551-2

Logos Verlag Berlin GmbH
Comeniushof, Gubener Str. 47,
10243 Berlin
Tel.: +49 (0)30 42 85 10 90
Fax: +49 (0)30 42 85 10 92
INTERNET: http://www.logos-verlag.de

Dedicated to my parents

Acknowledgements

This thesis summarizes the several years of work accomplished under the supervision of my doctoral advisor (*Doktorvater* in German) Prof. Dr.-Ing. habil. Clemens Merten at the Institute of Chemical Process Engineering (ICVT), University of Stuttgart with the financial assistance of German Academic Exchange Service (DAAD) and Higher Education Commission Pakistan.

First, I would like to express my deep thanks and kind gratitude to Prof. Merten who gave me both the freedom and the necessary guidelines and directions to develop and pursue different creative ideas. I am highly obliged to him for his permanent interest and continuous support of my work and for his always open doors for all the fruitful discussions. Without his help, support, advice and motivation, this work would not have been possible.

I owe great thanks to Dr. Aleysa for his active participation in many valuable discussions and for giving me the access to use the state of the art research facilities for completing the experimental part of this thesis under his research group at Fraunhofer-IBP Stuttgart. I am also thankful to Prof. Dr.-Ing. Philip Leistner for writing the co-principal report (*Mitbericht)* on this thesis and Prof. Dr.-Ing. Ulrich Nieken for chairing the doctoral examination.

I appreciate the very nice working environment at ICVT during all these years. I want to acknowledge Prof. Nieken, Manuel Huber and Manuel Hirschler for the technical support of running the OpenFOAM simulations on the computing cluster. I want to admire Manuel Hirschler for helping me in the translation of Abstract of this thesis into German. I also want to thank Hannes, Winfried, Philip, Wasim, Simone, Hendrik, Jeremias and Sebastian for the inspiration, motivation and good quality time.

I want to avail this opportunity to thank Jürgen and Oliver for the support during experiments and Ferdinand, Jacob, Niro, Martin and Valentin for the encouragement and all the good times during my visiting research stay at Fraunhofer-IBP. I want to recognize Olga and Bing of Institute of Building Energetics, University of Stuttgart for sharing the knowledge and useful information about their CFD investigations.

Finally, I want to thank my loving and caring parents back in home for their patience of finding me unavailable during my research stay in Germany and for their moral support and belief in the completion of this work.

Table of contents

Nomenclature

Latin symbols

C	Concentration	kg m^{-3}	-
C_μ	k-ε turbulence model constant	-	0.09
$C_{\varepsilon 1}$	k-ε turbulence model constant	-	1.44
$C_{\varepsilon 2}$	k-ε turbulence model constant	-	1.92
C_3	Dissipation coefficient	-	1.0
d	distance or length	m	-
D	Kinematic diffusivity	m^2 s^{-1}	-
g	Gravitational acceleration	m^2 s^{-1}	9.81
k	Turbulence kinetic energy per unit mass	m^2 s^{-2}	-
p	Pressure	Pa	-
p'	Modified pressure	Pa	-
P_k	Shear production of turbulence	kg m^{-1} s^{-3}	-
$S_{M\,S}$	Mass source	kg m^{-3} s^{-1}	-
S_M	Momentum source	kg m^{-2} s^{-2}	-
Sc_t	Turbulent Schmidt number	-	-
t	Time	s	-
U	Velocity	m s^{-1}	-

Nomenclature

x, y, z	Coordinates locations	m	-
Y	Mass fraction	-	-

Greek symbols

ε	Turbulence dissipation rate	$m^2\,s^{-3}$	-
Γ	Dynamic diffusivity	$kg\,m^{-1}\,s^{-1}$	-
Γ_t	Turbulent diffusivity	$kg\,m^{-1}\,s^{-1}$	-
ν	Kinematic viscosity	$m^2\,s^{-1}$	-
μ	Dynamic viscosity	$kg\,m^{-1}\,s^{-1}$	-
μ_{eff}	Effective viscosity	$kg\,m^{-1}\,s^{-1}$	-
μ_t	Turbulent viscosity	$kg\,m^{-1}\,s^{-1}$	-
ρ	Density	$kg\,m^{-3}$	-
σ_k	Turbulence model constant for the k equation	-	1.0
σ_ε	k-ε turbulence model constant	-	1.3
σ_ρ	Full buoyancy model constant	-	1.0

Subscripts

diff	Difference
eff	Effective
exp	Measured
i, j	Species of components
M	Source
ref	Reference

sim	Simulated
t	Turbulent
x, y, z	Coordinates locations

Abbreviations

2D	Two dimensional
3D	Three dimensional
ACH	Air changes per hour
BC	Boundary condition
BSL	Baseline Model
C++	a general-purpose programming language
CAD	Computer Aided Design
CEL	CFX Expression Language
CETIEB	Cost Effective Tools for Better Indoor Environment in Retrofitted Energy Efficient Buildings
CFD	Computational Fluid Dynamics
CPU	Central Processing Unit
DN	Diameter Nominal
EARSM	Explicit Algebraic Reynolds Stress Model
EPA	Environmental Protection Agency
FID	Flame Ionization Detector
IDLH	Immediately Dangerous to Life or Health
IR	Infrared

Nomenclature

LEL	Lower Explosive Limit
LES	Large Eddy Simulation
MSG05	March 2005 Madison Square Garden
NDIR	Nondispersive Infrared
NIOSH	National Institute for Occupational Safety and Health
PID	Photo Ionization Detector
ppm	Parts per million
RANS	Reynolds Averaged Navier Stokes
RNG	Re-Normalization Group
SIMPLE	Semi-Implicit Method for Pressure-Linked Equations
SST	Shear Stress Transport
UAV	Unmanned Aerial Vehicle
VOC	Volatile Organic Compound

Software programs

ANSYS CFX	a Computational Fluid Dynamics software
ANSYS CFD-Post	a software program for the post-processing
ANSYS DesignModeler	a Computer Aided Design software program
ANSYS FLUENT	a Computational Fluid Dynamics software
ANSYS Meshing	a mesh generation software program
ANSYS Workbench	a software platform
CFD-Urban	a Computational Fluid Dynamics software
Code_Saturne	an open source Computational Fluid Dynamics code

COMSOL Multiphysics	an engineering simulation software
EXACT	a simulation software
FEFLO-Urban	a simulation software
FEM3MP	a simulation software
FLACS	FLame ACceleration Simulator (a commercial Computational Fluid Dynamics Software)
IESVE Microflo	a simulation software
Mfix	a simulation software
Nek5000	a simulation software
OpenFOAM	an open source Computational Fluid Dynamics code
ParaView	a software program for the post-processing
Star-CCM+	a Computational Fluid Dynamics software
Star-CD	a Computational Fluid Dynamics software
SU^2	an open source simulation code

Chemical Formulae

CHOH	Formaldehyde
CH_4	Methane
C_3H_8	Propane
CO	Carbon monoxide
CO_2	Carbon dioxide
H_2	Hydrogen
NO	Nitric oxide

Nomenclature

N$_2$O   Nitrous oxide

SF$_6$ Sulfur hexafluoride

SO$_2$ Sulfur dioxide

Abstract

There are many sources of emissions produced by burning fuel for power or heat, through chemical reactions, and from leaks from industrial processes or equipment. There is always a possibility of a potential hazard when these gases enter into the indoor environment with the air flow. In order to prevent from gas suffocation, explosion and fire accidents, there is a need to observe and understand the different phenomena and the time-dependent spread, propagation and distribution of the leaked gas (e.g. propane) in the presence of air flow inside a room. The determination of the concentration profiles are necessary to evaluate the potential hazard posed by the gas spread. There is also a need to consider the option of performing numerical simulations because experiments can be quite expensive and time-consuming for further detailed studies. Therefore, this work has introduced a continuous measurement system as well as a simulation model based on the application of CFD to determine the spatial and temporal distribution of gaseous emissions with the air flow in the indoor environment.

The main objectives of this work are to develop an appropriate measurement methodology and a 3D CFD transient multicomponent numerical simulation model for the determination of spatial and temporal distribution of gaseous emissions with the air flow in the indoor environment. This work is also aimed at comparing the numerical simulation results of different CFD programs for a 2D base case model of indoor air flow with and without emission source under laminar and turbulent flow conditions for the purpose of developing a better basic understanding of the physical phenomena and for the selection of the suitable and appropriate CFD program for the further development of the simulation model. The ultimate goal is to develop and validate the numerical simulation model based on double configuration for the determination of gaseous emissions in the indoor environment which could be used as an effective and adequate tool for the loss prevention and risk mitigation in the industrial indoor environment in case of any accident or emergency situation and for the design of a safer indoor environment. One of the goals is also to apply the developed simulation model to the loss prevention and risk mitigation in the indoor environment and to study the influence of different parameters on the distribution of gaseous emissions in the presence of air flow in the indoor environment so as to minimize the expensive and time consuming experimentation efforts.

An experimental setup is developed, the measurement technique is selected and several experimental runs are conducted first involving the two components (propane and air) and then the three components (propane, CO_2 and air) for the purpose of developing a better understanding of the spread and distribution of the gaseous emissions in the room with the air flow and for the experimental validation of the developed 3D numerical simulation model based on double configuration. The flame ionization detectors and infrared photometers are employed to determine the time-dependent concentrations of propane and CO_2 respectively at different locations, whereas the flowrate of the gases is measured with the help of Wilson flow grid. The experiments are carried out with and without flow obstruction present inside the room.

As the preliminary investigations for the purpose of developing a better understanding of the physical phenomena and for the selection of the suitable and appropriate CFD program for the development of the 3D transient multicomponent simulation model based on the mechanism double configuration, it is found out that all the employed CFD codes (CFX, FLUENT and OpenFOAM) in the present work have almost the equal capabilities to model the indoor air flow, and the simulation results of the programs are in good agreement against each other for both the laminar flow and the turbulent flow. Moreover, the comparison of the numerical simulation results and the relative capabilities of CFX and FLUENT to model and determine the spatial distribution of formaldehyde emission in the presence of air flow under both the laminar flow condition and the turbulent flow condition in the room shows that the results from both the CFD programs are in good agreement with each other. However, CFX is selected for the development of the 3D transient multicomponent numerical simulation model based on the double configuration and for the further simulation studies because of its availability for this research work and the ease of implementation of the double configuration mechanism and the complex boundary conditions with the user defined expressions and functions.

The 3D transient multicomponent numerical simulation model to determine the spatial and temporal distribution of gaseous emissions inside the room with air flow is developed based on the mechanism of double configuration using ANSYS CFX 15.0 in which the steady state and transient analyses are sequenced in such a way that the transient simulation automatically starts upon the completion of the steady state simulation, whereby the results from the steady-state analysis are used as the initial guess for the transient analysis.

The validation of the developed 3D multicomponent numerical simulation model based on the mechanism of double configuration for the two components system and the three components system is done against the experimental results. The simulation results are found to be in good agreement against the experimental data. The difference between the simulation and the experimental results is found to be not more than 20 percent which is an acceptable standard for the CFD validations. The developed simulation model in this work is validated against the experimental data for both the cases with and without flow obstruction placed inside the room. It is also demonstrated that the developed simulation model could be used as an effective and adequate tool for the loss prevention and risk mitigation in the indoor environment in case of accident or emergency situation. It is also exhibited that there are significant influences of the different parameters on the distribution of pollutant in the presence of air flow inside the room.

The development of the experimental methodology and the simulation model for the simultaneous determination of gaseous and particulate emissions at different locations inside the room with the air flow is considered as an important future work. In this work, only one flow obstruction was placed inside the room. To determine the distribution of pollutants inside the room with the air flow under the more realistic situations, it is recommended to place more number of flow obstructions inside the room. The developed simulation model in this work is based on isothermal conditions. It is proposed to extend the developed simulation model to accommodate the non-isothermal conditions. The developed simulation model based on double configuration in CFX is suggested to be regenerated in OpenFOAM as the use of the open source CFD codes should be encouraged. The developed simulation model based on the double configuration has a potential to be employed for the design of the clean rooms and further scenarios of the loss prevention and risk mitigation in the industries. It is also suggested to employ the modeling concept developed in the present work to simulate the distribution of VOCs emissions in cars. Moreover, the development of a measurement methodology to measure the temporal and spatial concentration distribution of VOCs inside the vehicle cabin under different conditions is also seen as a potential future research work.

Abstract

Kurzfassung

Es gibt viele Emissionsquellen, die durch Verbrennen von Brennstoff für Energie oder Wärme, durch chemische Reaktionen und durch Leckagen von industriellen Prozessen oder Geräten erzeugt werden. Es besteht immer die Möglichkeit einer Gefährdung, wenn diese Gase mit der Luftströmung in die Innenumgebung gelangen. Zur Verhütung von Gasnebel-, Explosions- und Brandunfällen müssen die unterschiedlichen Phänomene und die zeitabhängige Ausbreitung und Verteilung des Leckgases (z.b. Propan) bei Vorhandensein von Luftströmen in einer Atmosphäre beobachtet und verstanden werden. Die Bestimmung der Konzentrationsprofile ist notwendig, um die potenzielle Gefährdung der Gaskonzentration zu bewerten. Es bedarf auch der Notwendigkeit, numerische Simulationen durchzuführen, da Experimente für weitere Detailstudien recht teuer und zeitraubend sein können. In dieser Arbeit wird daher ein kontinuierliches Messsystem sowie ein Simulationsmodell auf Basis der Anwendung von CFD zur Bestimmung der räumlichen und zeitlichen Verteilung von gasförmigen Emissionen mit dem Luftstrom im Innenraum eingeführt.

Die Hauptziele dieser Arbeit sind die Entwicklung einer geeigneten Messmethodik und eines 3D-CFD-Transienten-Multikomponenten-Simulationsmodells zur Bestimmung der räumlichen und zeitlichen Verteilung gasförmiger Emissionen mit dem Luftstrom im Innenraum. Diese Arbeit zielt auch darauf ab, die numerischen Simulationsergebnisse verschiedener CFD - Programme für ein 2D - Basisfallmodell des Innenluftstroms mit und ohne Emissionsquelle unter laminaren und turbulenten Strömungsverhältnissen zu vergleichen, um ein besseres Grundverständnis der physikalischen Phänomene zu erhalten und eine Auswahl des geeigneten CFD-Programms für die Weiterentwicklung des Simulationsmodells festzulegen. Ziel ist es, das numerische Simulationsmodell auf der Basis einer Doppelkonfiguration für die Bestimmung gasförmiger Emissionen im Innenraum zu entwickeln und zu validieren, die als effektives und adäquates Werkzeug für die Schadenverhütung und Risikominderung im industriellen Indoor-Umfeld zur Vermeidung von Unfall- oder Notsituation und für die Gestaltung eines sichereren Innenraums genutzt werden können. Eines der Ziele ist es, das entwickelte Simulationsmodell auf die Schadensverhütung und Risikominderung im Innenraum zu übertragen und den Einfluss verschiedener Parameter auf die Verteilung von gasförmigen Emissionen in Gegenwart von Luftströmen im Innenraum zu untersuchen. Damit werden teure und zeitraubende Experimente verringert.

Es wird ein experimenteller Aufbau entwickelt, die Messtechnik ausgewählt und mehrere experimentelle Versuche durchgeführt, wobei zunächst die beiden Komponenten (Propan und Luft) und dann die drei Komponenten (Propan, CO_2 und Luft) zum besseren Verständnis der Ausbreitung der gasförmigen Emissionen im Raum mit dem Luftstrom verwendet werden. Außerdem wird es verwendet um die experimentelle Validierung des entwickelten 3D numerischen Simulationsmodells durchzuführen. Zur Bestimmung der zeitabhängigen Konzentrationen von Propan bzw. CO_2 an verschiedenen Stellen werden Flammenionisationsdetektoren und Infrarot-Photometer eingesetzt, während die Durchflussmenge der Gase mit Hilfe des Wilson-Strömungsrasters gemessen wird. Die Experimente werden mit und ohne Strömungshindernis im Raum durchgeführt.

Als Voruntersuchungen zur Erarbeitung eines besseren Verständnisses der physikalischen Phänomene und zur Auswahl des geeigneten CFD-Programms für die Entwicklung des 3D-transienten Mehrkomponenten-Simulationsmodells auf der Grundlage der Mechanismus-Doppelkonfiguration ergibt sich, dass alle verwendeten CFD-Codes (CFX, FLUENT und OpenFOAM) in der vorliegenden Arbeit nahezu die gleichen Möglichkeiten besitzen, den Innenluftstrom zu modellieren, und die Simulationsergebnisse der Programme sowohl für die laminare Strömung als auch für die turbulente Strömung gut übereinstimmen . Darüber hinaus zeigt der Vergleich der numerischen Simulationsergebnisse und der Möglichkeiten von CFX und FLUENT zur Modellierung und Bestimmung der räumlichen Verteilung der Formaldehydemission in Gegenwart von Luftströmung, dass sowohl unter dem laminaren Strömungszustand als auch im turbulenten Strömungszustand im Raum die Ergebnisse der beiden CFD-Programme gut übereinstimmen. CFX wird jedoch für die Entwicklung des 3D-transienten Mehrkomponenten-Simulationsmodells auf Basis der Doppelkonfiguration und für die weiteren Simulationsstudien aufgrund seiner Verfügbarkeit für diese Forschungsarbeit und der leichten Implementierung des Doppelkonfigurationsmechanismus und der komplexen Randbedingungen ausgewählt. Dazu werden benutzerdefinierte Ausdrücken und Funktionen verwendet.

Das 3D-Transienten-Multikomponenten-Simulationsmodell zur Bestimmung der räumlichen und zeitlichen Verteilung gasförmiger Emissionen im Raum mit Luftströmung wird basierend auf dem Mechanismus der Doppelkonfiguration mit ANSYS CFX 15.0 entwickelt, bei dem die stationären und transienten Analysen so sequenziert werden, dass die transiente Simulation nach Erreichen des Steady-States automatisch startet, wobei die Ergebnisse der Steady-State-Analyse als anfängliche Vermutung für die Transientenanalyse verwendet werden.

Die Validierung des entwickelten 3D-Mehrkomponenten-numerischen Simulationsmodells, basierend auf dem Mechanismus der Doppelkonfiguration für das Zwei-Komponenten-System und das Drei-Komponenten-System, wird gegen die experimentellen Ergebnisse durchgeführt. Die Simulationsergebnisse liegen in guter Übereinstimmung mit den experimentellen Daten. Der Unterschied zwischen der Simulation und den experimentellen Ergebnissen liegt bei nicht mehr als 20 Prozent, was ein akzeptabler Standard für die CFD-Validierungen ist. Das entwickelte Simulationsmodell in dieser Arbeit wird gegenüber den experimentellen Daten für die Fälle mit und ohne Strömungshindernis im Raum validiert. Es wird auch gezeigt, dass das entwickelte Simulationsmodell als wirksames und adäquates Instrument zur Schadenprävention und Risikominimierung im Innenraum bei Unfall oder Notsituationen genutzt werden kann. Es wird auch gezeigt, dass es signifikante Einflüsse der verschiedenen Parameter auf die Verteilung von Schadstoffen in der Gegenwart von Luftströmung innerhalb des Raumes gibt.

Die Entwicklung einer experimentellen Methodik und eines Simulationsmodells zur gleichzeitigen Bestimmung von Gas- und Partikelemissionen an verschiedenen Stellen im Raum mit dem Luftstrom gilt als wichtige Zukunftsarbeit. In dieser Arbeit wurde nur ein Strömungshindernis innerhalb des Raumes platziert. Um die Verteilung der Schadstoffe innerhalb des Raumes mit dem Luftstrom unter realistischeren Bedingungen zu ermitteln, empfiehlt es sich, mehr Raumströmungshindernisse im Raum zu platzieren. Das entwickelte Simulationsmodell basiert auf isothermen Bedingungen. Es wird vorgeschlagen, das entwickelte Simulationsmodell auf die nicht-isothermen Bedingungen auszudehnen. Das entwickelte Simulationsmodell, das auf einer Doppelkonfiguration in CFX basiert, sollte in OpenFOAM implementiert werden, da die Verwendung der Open-Source-CFD-Codes gefördert werden sollte. Das entwickelte Simulationsmodell auf Basis der Doppelkonfiguration hat ein Potential für die Auslegung der Reinräume und weitere Szenarien der Schadensverhütung und Risikominderung in den Branchen. Es wird auch vorgeschlagen, das in der vorliegenden Arbeit entwickelte Modellierungskonzept zu verwenden, um die Verteilung der VOC-Emissionen in Autos zu simulieren. Darüber hinaus wird die Entwicklung einer Messmethodik zur Messung der zeitlichen und räumlichen Konzentrationsverteilung von VOCs innerhalb der Fahrzeugkabine unter verschiedenen Bedingungen auch als mögliche zukünftige Forschungsarbeit betrachtet.

Kurzfassung

Chapter 1   Introduction

1.1   Motivation

Gaseous dispersion

There are many sources of emissions produced by burning fuel for power or heat, through chemical reactions, and from leaks from industrial processes or equipment. There is always a possibility of a potential hazard when these gases enter into the indoor environment with the air flow. The suffocation, fire and explosion in the houses and indoor industrial environments induced by gas leakages can cause the human and property losses. One of the primary reasons for the fire and explosion in the houses and indoor industrial environments is the leaked gas concentrated within the explosive limits and when an ignition source is offered. Many people are being injured and killed due to either explosion or fire or suffocation caused by gas leakage in homes and in industrial environments. The characteristics of some reported incidents taken place due to gas leakage in houses in the different regions (India, Pakistan and USA) in 2013 – 2015 are shown in Table 1.1.

Many accidents have also been reported to occur due to toxic gas leakage in industrial environments. The characteristics of some reported accidents taken place due to gas leakage in industrial environments in 2011 – 2015 are shown in Table 1.2 [1].

Quite recently according to a news report of 28.08.2015 [2], a total of 18 people were injured in an ammonia gas leak accident in a gingerbread factory in southern German state of Bavaria. About 200 liter ammonia gas was leaked out from the gingerbread factory in Nuremberg. As a result, factory buildings needed to be evacuated. Altogether 150 workers were required to leave. Eighteen people suffered respiratory tract irritation, of whom two were so badly injured that they had to be sent to hospital. According to a news report of 23.10.2014 [3], a dramatic incident in the west of the country left one person dead and 10 more injured. The blast reportedly took place at a building site in the city of Ludwigshafen, Rhineland-Palatinate. It was believed to have been caused by a gas leak that caught alight.

Table 1.1:   Characteristics of some reported incidents taken place due to gas leakage in houses in different regions (India, Pakistan and USA) in 2013 – 2015

	Incident reporting date	Country	Cause of incident	Injuries / casualties / consequences
1.	13.06.2015 [4]	Ludhiana India	Ammonia leakage from a tanker	Six persons killed, 100 complained of breathing problem
2.	16.04.2015 [5]	India	Gas leakage from cylinder in a house followed by fire	Five persons injured
3.	03.03.2015 [6]	Quetta Pakistan	Explosions caused by gas leakage in different houses	Three people killed, three people injured
4.	24.02.2015 [7]	New Jersey USA	Natural gas leakage in a house followed by explosion	Five people killed/injured, 75 to 100 families evacuated and the gas line shut off
5.	22.02.2015 [8]	Lahore Pakistan	Toxic gas leakage from a generator in the basement	30 school children found faint in school
6.	11.01.2015 [9]	Texas USA	Propane leakage in a house followed by explosion	One woman killed, one kid injured
7.	05.12.2014 [10]	Gujranwala Pakistan	Gas leakage followed by explosion	One person killed, three persons injured
8.	05.12.2014 [10]	Mardan Pakistan	Suffocation due to gas leakage in house	Two persons killed
9.	03.06.2014 [11]	India	LPG cylinder leakage in a hotel followed by fire	Three people injured
10.	03.04.2014 [12]	India	LPG cylinder leakage followed by fire	One man killed, one woman injured
11.	12.11.2013 [13]	Rawalpindi Pakistan	Gas leakage and explosion in a house	Five people injured
12.	01.03.2013 [14]	Lahore Pakistan	Gas leakage in basement	Nine persons found faint in the basement

2

Table 1.2:   Characteristics of some reported accidents taken place due to gas leakage in industrial environments in 2011 – 2015 [1]

	Accident reporting date	Information	Country
1.	01.06.2015	Seven dead, one missing after Central China gas leak	China
2.	09.09.2014	NW China chemical plant ammonia leak poisons 33 people	China
3.	17.06.2014	Gas leakage kills two engineers at Vizag Steel Plant	India
4.	13.06.2014	6 dead, over 30 injured in gas leakage at Bhilai Steel Plant	India
5.	11.02.2014	20 women ill after ammonia leak in Mumbai	India
6.	13.08.2013	Toxic gas kills 4 factory workers	India
7.	02.05.2013	50 ill after gas leak from German chemical plant in Indore	India
8.	05.01.2013	2 died in gas leak at Hindustan Zinc plant	India
9.	21.09.2011	Gas leak in Tarapur factory kills 4 workers	India
10.	23.08.2011	35 hospitalised after gas leak, major tragedy averted	India
11.	03.08.2011	3 die in gas leak at Jindal steel plant	India
12.	24.01.2011	Ammonia leak at Dum Dum, 5,000 affected	India

In order to prevent from gas suffocation, explosion and fire accidents, there is a need to observe and understand the different phenomena and the time-dependent spread, propagation and distribution of the leaked gas (e.g. propane) in the presence of air flow inside a room. Based on the literature review shown in the next chapter, it has been found out that the full scale experimental and detailed numerical investigations for the purpose of development of better physical understanding of the time-dependent spread, propagation and distribution of the pollutant (e.g. propane) with the air flow with and without flow obstruction inside the room and for the purpose of validation of the numerical simulation model have seldom been realized. The determination of the concentration profiles are necessary to evaluate the potential hazard posed by the gas spread. There is also a need to consider the option of performing numerical simulations because experiments can be quite expensive and time-consuming for the further detailed studies. Therefore, this work should introduce a continuous measurement system as well as a simulation model to determine the spatial and temporal distribution of gaseous emissions with the air flow in the indoor environment.

VOCs emissions

The volatile organic compounds (e.g. benzene, formaldehyde, methylene chloride, hexane, toluene, trichloroethane, styrene, etc.), which can cause allergy, asthma, cough, headache, neurological and even cancer diseases, emit from different materials (like carpets, seats, paint, cleaning substances, lubricants, ventilation systems, plastics, stacks, reactors etc.) in different environments like buildings, car cabins, chemical plants etc. Indoor environments may be contaminated with about 100 to 1000 different VOCs which cause potential hazards to human. The sources of VOCs inside room buildings and interior of cars may be due to carpet, seats, paint, cleaning substances, lubricants, ventilation systems, plastics, interaction with the external environment etc. Around 275 organic chemicals have been identified, using experimental methods, in domestically manufactured cars in Japan. Similarly there are a number of different VOCs that are present in the surroundings of chemical plants and other establishments [15, 16, 17, 18]. Based on the literature review shown in the next chapter, the comparison of simulation results of different CFD programs for an application of indoor air flow with and without emission source inside the room under laminar and turbulent flow conditions has not been done so far. The emission source has not been specified by using a source term by any of the studies mentioned in the next chapter except Spengler [19, 20, 21, 22]. But the model developed by Spengler is based on laminar flow model. The developed simulation model in the present work is based on RANS turbulence modeling because of higher flow rate of air inside the room.

1.2 Objectives

The main objective of this work is to develop a suitable measurement methodology and a 3D transient multicomponent numerical simulation model based on the application of Computational Fluid Dynamics for the determination of spatial and temporal distribution of gaseous emissions with the air flow in the indoor environment. This work is also aimed at comparing the numerical simulation results of different CFD programs for a 2D base case model of indoor air flow with and without emission source under laminar and turbulent flow conditions for the purpose of developing a better basic physical understanding and for the selection of the suitable and appropriate CFD program for the development of the simulation model enabling the determination of the temporal and spatial distribution of gaseous pollutants with the air flow inside the room. The ultimate goal is to develop and validate the numerical simulation model for the determination of gaseous emissions in the indoor environment which could be used as an effective and adequate tool for the loss prevention and risk mitigation in the industrial indoor environment in case of any accident or emergency situation and for the design of a safer indoor environment. One of the goals is also to apply the developed simulation model to the loss prevention and risk mitigation in the indoor environment and to study the influence of different parameters on the concentration distribution of gaseous pollutants in the presence of air flow in the indoor environment to minimize the expensive and time consuming experimentation efforts.

1.3   Research tasks and thesis outline

Chapter 2 describes in detail the current state of research and technology for the experimental and numerical investigations of the distribution of emissions in the indoor environment. Moreover, the literature review of numerical modeling and simulation of indoor air flow is also shown. In addition, the literature survey of the studies related to the comparison of simulation results of different CFD programs for the different applications is also given. The inference from the literature review is exhibited and the gaps present in the current state of research and technology are highlighted. The contribution of the present work to the current state of research and technology is also mentioned.

The development of the 3D transient multicomponent numerical simulation model based on the mechanism of double configuration using Computational Fluid Dynamics to determine the spatial and temporal distribution of gaseous emissions with the air flow in the room is described in **Chapter 3**. In double configuration, the steady state and transient analyses are sequenced in such a way that the transient simulation automatically starts upon the completion of the steady state simulation and the results from the steady-state analysis are used as the initial guess for the transient analysis. In **Chapter 3**, the CAD modeling and meshing of the 2D base case model and the 3D developed model are demonstrated. The governing equations, the implemented boundary conditions and model parameters are explained. The method of simulation and the description of the different codes used in this work are mentioned.

Chapter 4 shows the comparison of the numerical simulation results of different CFD programs for a 2D base case model of indoor air flow with and without emission source under both laminar and turbulent flow conditions for the purpose of developing a better basic physical understanding and for the purpose of selection of the suitable and appropriate CFD program for the development of the 3D transient multicomponent simulation model based on the mechanism double configuration which enables the time-dependent determination of the spatial distribution of gaseous emissions with the air flow inside the room. This chapter concludes with the implementation of the developed simulation model using ANSYS CFX 15.0.

Chapter 5 deals with the detailed experimental determination of the gaseous emissions with the air flow in the room with and without flow obstruction for the purpose of developing a better understanding of the distribution of the gaseous emissions and the experimental validation of the developed simulation model. The full scale experimental setup, measurement technique, experimental procedure and conditions are described. The experimental results of the two components system and the three components system are then shown and discussed.

Chapter 6 shows the experimental validation of the developed 3D multicomponent numerical simulation model based on the mechanism of double configuration presented in Chapter 3 for the two components system (with and without flow obstruction) and the three components system. In addition, the comparison of the numerical simulation results of the three CFD programs with the existing experimental data to determine the velocity distribution inside the room (2D base case model) is also demonstrated.

The application of the developed simulation model to the loss prevention and risk mitigation in the indoor environment and to investigate the influence of different parameters on the concentration distribution of gaseous emissions in the presence of air flow in the indoor environment is exhibited and discussed in **Chapter 7**.

The **last chapter** summarizes the present work and shows room for the improvement of the developed experimental methodology and simulation model for the determination of the distribution of gaseous emissions with the air flow in the room. It also provides an outlook on the further possibilities and prospects of using the developed measurement methodology and simulation model.

Chapter 2   State of research and technology

2.1   Indoor air flow

Air flow is considered to be one of the most important parameters in the spread and distribution of the gaseous emissions in the indoor environment. The 3D CFD numerical simulation results with the laminar, k-ε turbulence, and RNG k-ε turbulence numerical models were compared with the experimental measurements of indoor air flows in a one-tenth sub-scale model room [23]. This study shows that the RNG model most accurately predicts the flow in a partitioned room, capturing the gross effects of a large flow obstruction. In another study [24], full-scale experimental and CFD methods were used to investigate the velocity and temperature fields in a mechanically ventilated enclosure in three cases of ventilation air temperature: an isothermal case, a hot case and a cold case. Four two equations turbulence models: a k-ε realizable model, a k-ε RNG model, a k-ω model and a k-ω SST model were tested against the experimental results. The models could predict reasonably the hot and isothermal cases global values of temperature and velocity, none of the models was reliable concerning the studied cold case (relatively lower air temperature). However, a detailed analysis of the jet showed that none of the models was able to predict the exact experimental velocity and temperature fields.

The most important factors that can influence the applications of CFD for studying air distribution are the selection of an appropriate CFD approach and a turbulence model. One paper [25] summarizes recent progress in CFD turbulence modeling and its application to some practical indoor environment studies and also further identifies a few turbulence models that show great potential for modeling airflows in enclosed environments. The review identified eight prevalent and/or recently proposed turbulence models for indoor airflow prediction. The standard k-ε model with wall functions is still widely used and provides acceptable results (especially for global flow and temperature patterns) with good computational economy. The model may have difficulty dealing with special room situations (e.g., high buoyancy effect and/or large temperature gradient). The RNG k-ε model (Yakhot and Orszag 1986) provides similar (or slightly better) results as the standard model and is also widely used for airflow simulations in enclosed environments. The LES model provides more detailed and, perhaps, more accurate predictions for indoor airflows, which could be important for understanding the flow mechanism. However, the high demand on computing time and user knowledge still makes LES a tool mainly for research and RANS model development. In another paper by the same authors [26], the evaluation and comparison of the selected eight turbulence models in terms of accuracy and computing cost for several indoor benchmark cases showed that the LES provides the most detailed flow features, while the computing time is much higher than for RANS models, and the accuracy may not always be the highest. Among the RANS models studied, the RNG k-ε model and a modified v^2-f model perform the best overall in four cases studied. Meanwhile, the other models have superior performance only in particular cases. While each turbulence model has good accuracy in certain flow categories, each flow type favors different turbulence models.

2.2 Multicomponents

Gaseous distribution

Wu et al. [27] conducted the simulation of the indoor gas (natural gas) leak by using CFD. The regions of concentrations within the explosive limits in the room were calculated. The influence by the release rate of gas leaks and the velocity of the outdoor air flows was studied. The CFD model FLACS was described and the examples of its application to indoor leaks of propane and natural gas to demonstrate the effect of leak location and momentum on the behavior of the flammable cloud were provided by Gavelli et al. [28]. This research also stated that the experiments tend to be time-consuming and expensive and, in the case of indoor releases of flammable gases, might pose significant risks.

Barley et. al. [29] described the buoyancy-driven passive ventilation of H_2 from buildings through vents to the outside with a goal to ascertain the relationship between vent design, leakage rate, maximum concentration, and other variables, which leads to design guidelines, an understanding of the limitations of this approach, and recommendations for codes and standards. The CFD model has not yet been compared with experimental results as a check on its accuracy and the research plan includes this work as a future activity. The CFD based model for indoor risk assessment considering accidental release of a sustained, small, undetected leak of a dense toxic gas (chlorine) in an indoor environment of chemical process industry was proposed by Siddiqui et al. [30]. Muharam et al. [31] simulated the gas (methane) leaks in a city gas utilization system (room) using COMSOL Multiphysics. The distribution of the diffusing gases in the room was calculated, and the times required to achieve the flammability limit were predicted. Furthermore, the simulation on the times required to evacuate leaking gas until it reaches a safe condition was also performed with various exhaust fan velocities.

In a more recent research by Nagaosa [32], three numerical runs of flammable gas spread were performed using two gases with different density as a series of case studies, and compared the effect of gas density on the three-dimensional spread of these gases in the residential space. The effect of the gas leakage rate on the spread of the gases leaked into the three-dimensional enclosed space was also demonstrated. This numerical modeling study did not introduce a turbulence model to compute the concentration profiles of the leaked gas. This study states that the implementation of a suitable turbulence model is necessary to reduce hardware requirements and CPU time. In addition, the results of the numerical predictions have not been validated by using a series of laboratory experiments. This study also recommends that the validations of the numerical results by laboratory and in-situ measurements should be provided in the future.

The numerical and experimental investigations of ventilation efficiency of different ventilation patterns arranged by two inlet and two outlet diffusers at different locations was performed by Chung et al. [33] . The tracer decay method was employed to measure the relative ventilation efficiency. A constant tracer gas (CO_2) addition in the supply was stopped after steady-state conditions had been reached and the subsequent decay of concentrations vs. time was recorded. The experimental results were used to validate the simulation results. The experimental and numerical investigations of tracer gas (SF_6) transport in an office and in an airliner cabin using RNG k-ε turbulence model have been done by Cheong et al. [34] and Zhang et al. [35] respectively. In addition, Zhang et al. [35] also measured and simulated the particle transport.

In a more recent study done by Bing et al. [36, 37], the non-isothermal quasi steady numerical simulations and experimental measurements were done in a full scale test chamber to study the concentration distribution of tracer gas Nitrous oxide (N_2O) under the project of *Cost Effective Tools for Better Indoor Environment in Retrofitted Energy Efficient Buildings*

(CETIEB). Moreover, the impact of a decontaminating surface on the contaminant distribution in a room with the ventilation was also considered. The tracer gas source was placed at the table in the room. The CFD software program used for this numerical investigation was ANSYS FLUENT. RNG k-ε turbulence model was used to simulate the air flow in the room. Temperatures and air velocity in the room were measured by thermocouples and heat loss anemometers respectively. Whereas, the concentration of Nitrous oxide (N_2O) was measured by NDIR gas analyzer. The experimental and simulation results were in good agreement.

Based on the literature review shown above, it has been found out that the full scale experimental and detailed numerical investigations for the purpose of development of better physical understanding of the time-dependent spread, propagation and distribution of the pollutant (e.g. propane) with the air flow with and without flow obstruction inside the room and for the purpose of validation of the numerical simulation model have seldom been realized.

Particulate emissions

The numerical simulation of contaminant particle transport in an enclosure was performed by Chung [38] using EXACT 3. No experimental validation for the particle transport was given. The experimental and numerical investigations of particle transport in the ventilated rooms using Lagrangian particle tracking method was performed by [39, 40, 41].

VOCs emissions

Murakami et al. [42, 43, 44] presented the physical models based on the fundamental physio-chemical diffusion and adsorption/desorption principles in numerically analyzing the transportation of VOCs from building materials in a room using CFD. The turbulence model used was low Reynolds number k-ε model. No experimental validation for the concentration distribution was performed. The numerical simulation using FLUENT/UNS was performed to predict the VOC (toluene) removal in a small scale painting process with the ventilation [45]. The steady state toluene concentration and transient ventilation characteristics of toluene were analyzed to predict room air flow pattern, air change rate and room mean age by employing standard k-ε turbulence model. No experimental validation was carried out. In another study done by Deng et al. [46], a new CFD model using the equivalent air-phase concentration was developed based on the general adsorption isotherm. The same authors have also implemented the newly developed model to investigate the VOCs emission from a new carpet in a room under different ventilation strategies using FLUENT 6 [47, 48]. The simulation model was validated against the measurement results taken from the environmental test chamber.

Spengler [19, 20, 21, 22] developed a measurement method to determine the spatial and temporal distribution of evaporative emissions inside the air-intake system of automobiles and to localize the sources and generation rates without affecting the physical processes e.g. diffusion. The PID gas sensors calibrated with the reference measurement method (Low-Flow-FID) were used to measure the volatile hydrocarbons inside the built laboratory geometries. In addition, a simulation method using ANSYS CFX was developed to simulate evaporation, diffusion and distribution of hydrocarbons inside the air-intake system. The validation of the simulation model was done against the experimental results.

The emission source has not been specified by using a source term by any of the studies mentioned above except Spengler [19, 20, 21, 22]. But the model developed by Spengler was based on laminar flow model. The developed simulation model in the present work is based on RANS turbulence modeling because of higher flow rate of air inside the room.

2.3 Comparison of simulation results of different CFD programs

The numerical simulations of the turbulent flow in an asymmetric two-dimensional diffuser carried out using three commercial CFD codes CFX, Fluent, and Star-CD have shown that the three employed codes show very similar characteristics in terms of convergence and accuracy [49]. The features of several CFD packages that could be coupled to an external electromagnetic code have been reviewed [50]. Five CFD models (CFD-Urban, FLACS, FEM3MP, FEFLO-Urban and FLUENT-EPA) have been applied to the same urban boundary layer scenario to enable model comparisons for the three-dimensional building database for Manhattan licensed by the Vexcel Corporation [51]. The simulations by the five models were qualitatively similar and they agreed fairly well with each other and with the MSG05 (March 2005 Madison Square Garden) flow observations.

The OpenFOAM simulation results were compared against the FLUENT results firstly for the boundary layer of a flat moving wall and secondly for the incompressible and isothermal and turbulent free-jet in air [52]. The results of OpenFOAM were very similar to those of FLU-ENT for the considered two cases. The numerical simulation of turbulent flow through both staggered and in-line tube bundle using CFX and FLUENT demonstrated that FLUENT showed the overall superior prediction performance as compared to CFX for the converged solution [53]. The comparison among CFD packages (Mfix, Nek5000 and OpenFOAM) for the purpose of choosing the suitable one for the modeling of gasification process revealed that Nek5000 was not applicable and OpenFOAM was more straightforward and there was access to a lot of options that were not available in Mfix [54].

The comparison of three CFD software packages FLUENT, Star-CCM+ and OpenFOAM on their ability to model the fluid phase in the submerged jet impingement test showed that FLUENT and OpenFOAM give very similar velocity profiles and capture the position of the peak velocity magnitude, whereas StarCCM+ is slightly different, mainly in the axial plane [55]. The comparison of three open source CFD codes SU^2, OpenFOAM, and Code_Saturne for the aerodynamic analysis of a mini-UAV (Unmanned Aerial Vehicle) has demonstrated that Code_Saturne and OpenFOAM gave the best results and the trends (as well as values) of drag and lift coefficients were very similar for OpenFOAM and Code_Saturne [56].

In a very recent study [57], FLUENT has been recommended for CFD modeling of compli-cated indoor fluid environment because of its best performance compared with Star-CCM+ and IESVE Microflo according to the comparison results of the three CFD programs for mod-eling of the baseline model of an ambulance hall.

It is important to note that the comparison of simulation results of different CFD programs for an application of indoor air flow with and without emission source inside the room under lam-inar and turbulent flow conditions has not been done by any of the contributions mentioned above.

2.4 Contribution of the present work

The present work focuses on the development of an experimental methodology and a numeri-cal simulation model to determine the distribution of gaseous emissions with the air flow in-side the room. The present work not only focuses on the experimental and numerical investi-gation of influence of obstruction on the concentration distribution of gaseous pollutant inside the room but also focuses on the in-depth and detailed studies of influence of different parameters on the distribution of gaseous emission with the air flow inside the room. In addition, the present work not only considers two components (propane and air) but also considers three components (propane, carbon dioxide and air). The developed simulation model in the present work is based on RANS turbulence modeling because of higher flow rate of air inside the room. Moreover, the simulation mechanism based on double configuration which reflects more close reality has been used. Also the user defined expressions and statements for the boundary conditions to realize the reality have been employed. The present work also considers the application of the developed and experimentally validated simulation model to the loss prevention and risk mitigation in the indoor environment in case of any accident or emergency situation.

Chapter 3   Numerical Modeling

This chapter describes the development of a 3D transient multicomponent numerical simulation model based on the double configuration for the determination of spatial and temporal distribution of gaseous emissions with the air flow in the indoor environment. Double configuration means that the steady state and transient analyses in the simulation model have been sequenced in such a way that the transient simulation automatically starts upon the completion of the steady state simulation, and the results from the steady-state analysis are used as the initial guess for the transient analysis. In this chapter, first the geometry creation and the mesh generation of the 2D base case model and the 3D test room are described. The governing equations are then explained followed by the descriptions of the double configuration and the implemented boundary conditions and model parameters. At the end of this chapter, the method of simulation is explained followed by the description of the different codes used in this work.

In the scope of this work, the base case model, for which the specifications of the geometry dimensions and the literature values for the velocity distribution are taken from the literature [58], has been employed to compare the numerical simulation results of different CFD programs for the indoor air flow with and without emission source under laminar and turbulent flow conditions for the purpose of developing a better basic physical understanding and for the purpose of selection of the suitable CFD program for the development of the 3D transient multicomponent simulation model based on the mechanism double configuration. The numerical simulation results with the base case model are demonstrated in Chapter 4. Moreover, in this work the test room has been employed to conduct the detailed numerical and experimental investigations of the pollutant spread and distribution inside the room with the air flow for the purpose of developing a better physical understanding and the experimental validation of the developed simulation model. The experimental investigation and the model validation results are shown in Chapter 5 and Chapter 6 respectively.

3.1   CAD modeling and meshing

The description of the CAD modeling and meshing of the base case model and the test room is explained in detail below.

3.1.1   Base case model

The geometry of the 2D model, for which the specifications of the geometry dimensions and the literature values for the velocity distribution are taken from the literature [58], is created using ANSYS DesignModeler 15.0 [59]. The detailed discussion about the description of this base case model is given in Chapter 4. The dimensions of this 2D model are described in Figure 3.1.

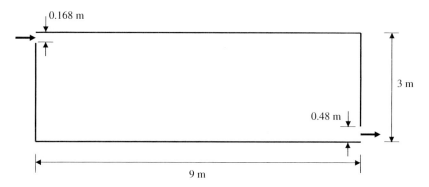

0.168 m

3 m

0.48 m

9 m

Figure 3.1:   Sketch of 2D base case model [58]

Only the flow domains have been drawn in which the equations of fluid flow will be solved. The inlet and outlet channels are extended so as to avoid the backflow. The lengths of the inlet channel and outlet channel are 1 m and 3 m respectively. As ANSYS CFX 15.0 [60] works on three dimensional models, a thin solid with a thickness of 0.02 m is produced and only one element row in the third spatial direction (z-direction) is placed. A two-dimensional model can directly be simulated in ANSYS FLUENT 15.0 [61]. The mesh of the base case model produced in ANSYS Meshing 15.0 [62] is shown in Figure 3.2. A mesh is produced with a total number of elements 11460 and the element size of 0.05 m. The meshing method used is "Hex Dominant". As the hex meshing is recommended for the relatively clean and simple geometry, the hex meshing is done because the base case model is simple and there are no complex mesh regions. Also, hex meshing leads to reduced element count and hence, the reduced run time. Moreover, the hex meshing leads to reduced numerical error.

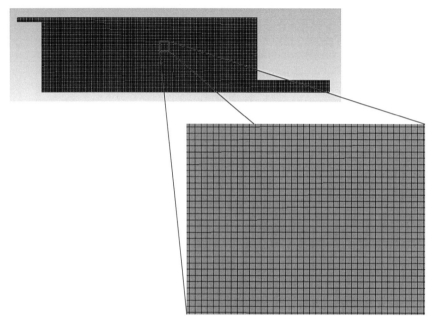

Figure 3.2:   Mesh of the 2D base case model

3.1.2   3D Test room

The 3D CAD model of the test room with all the main dimensions is shown in Figure 3.3. The detailed description of this test room is given in Chapter 5. The geometry of this 3D model is created with the help of ANSYS DesignModeler 15.0 [59]. The surface of the distributor is modeled as a non-porous surface so as to make the simulation model relatively simple. The tetrahedron mesh of this test room is generated using ANSYS Meshing 15.0 [62] and is exhibited in Figure 3.4.

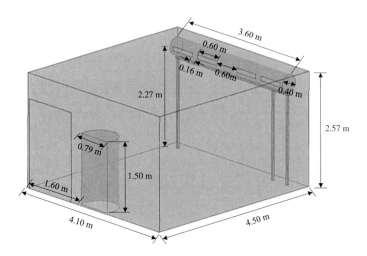

Figure 3.3:   3D CAD model of the test room with all the main dimensions

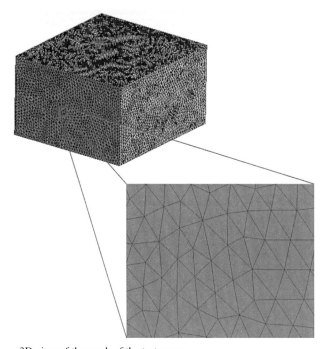

Figure 3.4:   3D view of the mesh of the test room

The tetrahedral meshing is recommended for meshing the complex regions. For this reason, tetrahedron mesh is generated because the test room contains some complex flow regions especially near the flow distributor and the exhaust section. The mesh independence study was conducted for the purpose of the selection of a better mesh, but has not been given in this written dissertation. The simulations with only the air flow were performed using the six tetrahedron meshes with the number of elements 57376, 85614, 202745, 296066, 430754 and 600085. The mesh with the 202745 number of elements and the minimum edge length of 0.00635 m was selected because the simulation results were become almost independent of the further increase in the number of elements, and the computation time was also not too much high as for the other meshes with the higher number of elements.

3.2   Governing equations

In this section, the governing equations for the modeling of distribution of gaseous emissions with the air flow inside the room are described. First, the modeling equations for the simulation of indoor air flow are presented because the air flow plays an important role for the spread, transportation and distribution of gaseous pollutant/s in the indoor environment. This is followed by the description of the equations for the multicomponent simulation of transportation and distribution of gaseous pollutants with air flow in the indoor environment.

3.2.1   Laminar flow

After calculating the Reynolds number from Eq. (3.1), the state of the flow is characterized and the corresponding simulation model (laminar or turbulent) is selected.

$$Re = \frac{\rho \cdot U \cdot d}{\mu} \tag{3.1}$$

When the Reynolds number is less than the critical value, the laminar flow is expected. The laminar flow is simulated by the solution of the unsteady Navier-Stokes equations [63]. For the incompressible flow, ρ = const.

The continuity equation is presented in Eq. (3.2)

$$\rho \frac{\partial U_x}{\partial x} + \rho \frac{\partial U_y}{\partial y} + \rho \frac{\partial U_z}{\partial z} = S_{M\,S} \tag{3.2}$$

Where, S_{MS} is the mass source term.

The momentum equations in x, y and z directions are shown in Eq. (3.3)

$$\left[\frac{\partial U_x}{\partial t} + U_x \frac{\partial U_x}{\partial x} + U_y \frac{\partial U_x}{\partial y} + U_z \frac{\partial U_x}{\partial z}\right] = -\frac{1}{\rho}\frac{\partial p}{\partial x} + v\left[\frac{\partial^2 U_x}{\partial x^2} + \frac{\partial^2 U_x}{\partial y^2} + \frac{\partial^2 U_x}{\partial z^2}\right]$$

$$\left[\frac{\partial U_y}{\partial t} + U_x \frac{\partial U_y}{\partial x} + U_y \frac{\partial U_y}{\partial y} + U_z \frac{\partial U_y}{\partial z}\right] = -\frac{1}{\rho}\frac{\partial p}{\partial y} + v\left[\frac{\partial^2 U_y}{\partial x^2} + \frac{\partial^2 U_y}{\partial y^2} + \frac{\partial^2 U_y}{\partial z^2}\right] \qquad (3.3)$$

$$\left[\frac{\partial U_z}{\partial t} + U_x \frac{\partial U_z}{\partial x} + U_y \frac{\partial U_z}{\partial y} + U_z \frac{\partial U_z}{\partial z}\right] = -\frac{1}{\rho}\frac{\partial p}{\partial z} + v\left[\frac{\partial^2 U_z}{\partial x^2} + \frac{\partial^2 U_z}{\partial y^2} + \frac{\partial^2 U_z}{\partial z^2}\right]$$

with the kinematic viscosity $v = \mu/\rho$.

Whereas, the concentration equation is exhibited in Eq. (3.4)

$$\left[\frac{\partial C}{\partial t} + U_x \frac{\partial C}{\partial x} + U_y \frac{\partial C}{\partial y} + U_z \frac{\partial C}{\partial z}\right] = D_{ij}\left[\frac{\partial^2 C}{\partial x^2} + \frac{\partial^2 C}{\partial y^2} + \frac{\partial^2 C}{\partial z^2}\right] \qquad (3.4)$$

3.2.2 Turbulent flow

The turbulent flow is expected when the Reynolds number is greater than the critical value. Turbulence models have been specifically developed to account for the effects of turbulence without recourse to a prohibitively fine mesh and direct numerical simulation [63]. In general, turbulence models seek to modify the original unsteady Navier-Stokes equations by the intro-duction of averaged and fluctuating quantities to produce the Reynolds Averaged Navier-Stokes (RANS) equations. Two equation turbulence models are widely used because they offer a good compromise between cost and numerical calculation accuracy [63, 64, 65]. Both the velocity and length scales are solved using separate transport equations, therefore these models are referred to as "two-equation model". The standard k-ε model belongs to this class and has become the standard model of practical technical flow calculations at the present time since it was suggested by Launder and Spalding [66]. The standard k-ε model is based on the transport equations for the turbulent kinetic energy k and its dissipation rate ε. k is the turbu-lence kinetic energy and is defined as the variance of the fluctuations in velocity. ε is the tur-bulence eddy dissipation (the rate at which the velocity fluctuations dissipate). After the Reynolds averaging, the continuity equation is given by Eq. (3.5):

$$\frac{\partial \rho}{\partial t} + \frac{\partial}{\partial x_j}(\rho U_j) = S_{M\,S} \tag{3.5}$$

Where, $S_{M\,S}$ is the mass source term.

And the momentum equation is given by Eq. (3.6):

$$\frac{\partial \rho U_i}{\partial t} + \frac{\partial}{\partial x_j}(\rho U_i U_j) = -\frac{\partial p'}{\partial x_i} + \frac{\partial}{\partial x_j}\left[\mu_{eff}\left(\frac{\partial U_i}{\partial x_j} + \frac{\partial U_j}{\partial x_i}\right)\right] + S_M \tag{3.6}$$

Wherein S_M is the sum of the body forces, μ_{eff} shown in Eq. (3.7) is the effective viscosity accounting for turbulence and p' presented in Eq. (3.8) is the modified pressure [64].

$$\mu_{eff} = \mu + \mu_t \tag{3.7}$$

$$p' = p + \frac{2}{3}\rho k + \frac{2}{3}\mu_{eff}\frac{\partial U_k}{\partial x_k} \tag{3.8}$$

where μ_t is the turbulence viscosity. In the k-ε model, the turbulence viscosity is linked to the turbulence kinetic energy and dissipation by the relationship shown in Eq. (3.9);

$$\mu_t = C_\mu \rho \frac{k^2}{\varepsilon} \tag{3.9}$$

C_μ is a constant and is typically equal to 0.09 [63, 64].

The values of k and ε are calculated directly from the differential transport equations for the turbulent kinetic energy and turbulent dissipation as shown in Eq. (3.10) and Eq. (3.11) respectively [63, 64, 65, 66].

$$\frac{\partial(\rho k)}{\partial t} + \frac{\partial}{\partial x_j}(\rho U_j k) = \frac{\partial}{\partial x_j}\left[\left(\mu + \frac{\mu_t}{\sigma_k}\right)\frac{\partial k}{\partial x_j}\right] + P_k - \rho\varepsilon + P_{kb} \tag{3.10}$$

$$\frac{\partial(\rho\varepsilon)}{\partial t} + \frac{\partial}{\partial x_j}(\rho U_j \varepsilon) = \frac{\partial}{\partial x_j}\left[\left(\mu + \frac{\mu_t}{\sigma_\varepsilon}\right)\frac{\partial\varepsilon}{\partial x_j}\right] + \frac{\varepsilon}{k}(C_{\varepsilon 1}P_k - C_{\varepsilon 2}\rho\varepsilon + C_{\varepsilon 1}P_{\varepsilon b}) \tag{3.11}$$

whereby $C_{\varepsilon 1}, C_{\varepsilon 2}, \sigma_k$ and σ_ε are constants and the values of these constants are 1.44, 1.92, 1.0 and 1.3 respectively [63, 64, 65, 66].

P_k is the turbulence production due to the viscous forces, which is modeled by Eq. (3.12)

$$P_k = \mu_t \left(\frac{\partial U_i}{\partial x_j} + \frac{\partial U_j}{\partial x_i} \right) \frac{\partial U_i}{\partial x_j} - \frac{2}{3} \frac{\partial U_k}{\partial x_k} \left(3\mu_t \frac{\partial U_k}{\partial x_k} + \rho k \right) \tag{3.12}$$

P_{kb} and $P_{\varepsilon b}$ represent the influence of the buoyancy forces and are described by the full buoyancy model [64, 65];

$$P_{kb} = -\frac{\mu_t}{\rho \sigma_\rho} g_i \frac{\partial \rho}{\partial x_i} \tag{3.13}$$

$$P_{\varepsilon b} = C_3 \cdot \max(0, P_{kb}) \tag{3.14}$$

where Turbulence Schmidt Number $\sigma_\rho = 1$ and Dissipation Coefficient $C_3 = 1$.

In nature the mass transfer occurs through convection and diffusion. For the turbulent flow, the spread, transportation and distribution of the pollutants with the air flow is considered to be a convection dominated process [64, 65] (also presented in Section B.2.2). The convection is determined by a pressure gradient. The convection currents transport the material along the streamlines. In addition to the macroscopic mass transfer, there still exists the interaction between different molecules which is referred to as "diffusion". The diffusion is due to the non-directional random movement of particles due to their thermal energy or non-uniform concentration. The Fick's law is used to describe the diffusion of a very dilute component in a carrier fluid with large quantities. For the multicomponent simulation, the scalar transport equations are solved for velocity, pressure and turbulence fields to model the bulk motion of the fluid. However, additional equations are solved to determine how the components of the fluid are transported within the fluid. The influence of the multiple components is felt only through property variation by virtue of differing properties for the various components. Each component has its' own equation for conservation of mass [64]. The Reynolds averaged transport equation for mass fraction is given by Eq. (3.15);

$$\frac{\partial \rho Y_i}{\partial t} + \frac{\partial}{\partial x_j} (\rho U_j Y_i) = \frac{\partial}{\partial x_j} \left[\left(\Gamma_i + \frac{\mu_t}{Sc_t} \right) \frac{\partial Y_i}{\partial x_j} \right] + S_{Y_i} \tag{3.15}$$

Where $\Gamma_i = \rho D_i$.

3.3   Boundary conditions

The boundary conditions are very important to fully define the flow simulation and for the better comparison of the simulation results with the experimental results. In this section, the general boundary conditions such as inlet, outlet, wall and symmetry implemented in the simulation models are discussed. The most robust combination of boundary conditions has been employed i.e. velocity/mass flow at an inlet and static pressure at an outlet [65].

Inlet

Inlets are used predominantly for the regions where inflow is expected [65]. For the base case model of indoor air flow, the magnitude of the inlet velocity of air is specified and the direction is taken to be normal to the boundary [64]. For the turbulent flow case, the values of the turbulence kinetic energy k and the turbulence eddy dissipation ε are taken from the literature [58]. For the case of formaldehyde emission, the amount of fluid introduced is specified as "Fluid Mass Source" on the subdomain with units kg m^{-3} s^{-1} in CFX [64, 65]. The detailed overview of different boundary conditions implementations in CFX, FLUENT and Open-FOAM for the turbulent flow and laminar flow has been illustrated in Chapter 4 and Chapter 6.

For the simulation model developed in the present work for the prediction of the pollutants spread, transportation and distribution with the air flow, the boundary mass flow rate (in kg s^{-1}) of air is specified and the flow direction is given as normal to the boundary which assumes that a uniform mass influx exists over the entire boundary [64, 65]. The mass source contribution to the conservation equation for each of the propane and CO_2 is specified as "Total Source" with units kg s^{-1}. CEL (CFX Expression Language) expressions and statements have been used to implement this boundary condition. Since there is exactly no idea about the inlet turbulence, the recommended option "Medium (Intensity = 5 %)" corresponding to 5 % intensity and a viscosity ratio of 10 has been used in CFX [65]. The detailed discussion of the different values of the boundary conditions and their implementation for the different cases is presented in Chapter 4, Chapter 6 and Chapter 7.

Outlet

The most commonly used option in CFX "Average Static Pressure" of 0 Pa "Over Whole Outlet", which allows the pressure profile at the outlet to vary based on upstream influences while constraining the average pressure to the user-specified value, has been used for the regions where outflow is expected [64, 65].

Wall

The wall boundary condition is used in the simulation models to the solid impermeable boundaries to fluid flow. The impact of the wall on the flow is implemented through a "No slip" condition where the velocity of the fluid at the wall boundary is set to zero and hence the boundary condition for the velocity is given by (3.16)

$$U_{wall} = 0 \tag{3.16}$$

The wall function approach has been employed to model the flow near the no-slip wall for the turbulent flow [63, 64, 65].

Symmetry plane

As ANSYS CFX works on three dimensional models, a thin solid with a thickness of 0.02 m is produced and only one element row in the third spatial direction (z-direction) is placed in the 2D base case model. The symmetry plane boundary condition is defined on the front and back planes which imposes constraints that 'mirror' the flow on either side of it. The normal velocity component at the symmetry plane boundary and the scalar variable gradients normal to the boundary are set to zero [64].

$$U_n = 0 \tag{3.17}$$

$$\frac{\partial \varphi}{\partial n} = 0 \tag{3.18}$$

3.4   Model parameters

The model parameters are very important for the better comparison of the simulation results with the experimental results. For the turbulent flow, the spread, transportation and distribution of the pollutants with the air flow is considered to be a convection dominated process, the kinematic diffusivity of the pollutant can have little effect (also presented in Appendix B.2.2). However, in order to accommodate both a convective and diffusive flow, the values of the kinematic diffusivities for both propane and carbon dioxide have been specified. The values of kinematic diffusivities of propane and carbon dioxide in air at 298 K (25 °C) and 101325 Pa (1 atm) are 0.000012 m^2 s^{-1} [67] and 0.000016 m^2 s^{-1} [68] respectively. For the pollutant spread, the ambient temperature and ambient pressure conditions have been assumed. The densities of propane, carbon dioxide and air at 298 K (25 °C) and 101325 Pa (1 atm) are 1.8316 kg m^{-3}, 1.8080 kg m^{-3} and 1.1800 kg m^{-3} respectively [69]. It is assumed that the components form an ideal mixture. The "variable composition mixture" and "ideal mixture" options are chosen in CFX [64, 65]. With these options, a mixture property is computed as a mass-weighted average of the component properties. The influence of gravity is implemented in the simulation model using the buoyancy approach [64, 65]. For this purpose, a source term $S_{M,buoy}$ given in Eq. (3.19) is added to the momentum equations.

$$S_{M,buoy} = (\rho - \rho_{ref})g \qquad (3.19)$$

The density of air (1.1800 kg m^{-3}) at 298 K (25 °C) and 101325 Pa (1 atm) is used as the reference density ρ_{ref}. The value of gravitational acceleration g is 9.81 m^2 s^{-1}.

3.5   Double configuration

The developed simulation model in the present work for the determination of the pollutants spread, transportation and distribution with the air flow inside the room is based on the double configuration in which the steady state and transient analyses have been sequenced in such a way that the transient simulation automatically starts upon the completion of the steady state simulation as shown in Figure 3.5. This is achieved by setting up the configurations of these analyses in a subsequent step by using the configuration control in ANSYS CFX 15.0.

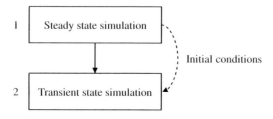

Figure 3.5:   Simulation mechanism based on the double configuration

The first configuration is a steady-state analysis involving only air flow and no pollutant/s inlet into the room through the distributor. The second configuration is a transient-state analysis involving pollutant/s inlet into the room along with the air through the distributor. The second configuration takes the setup of the steady state and adapts it for a transient analysis, hence the results from the steady-state analysis are used as the initial guess for the transient analysis. In simple words, the steady state and transient analyses have been sequenced in such a way that the transient simulation automatically starts upon the completion of the steady state simulation, and the results from the steady-state analysis are used as the initial guess for the transient analysis.

3.6   CFD procedure

CFD is a computer-based tool for the numerical investigation of systems involving fluid flow, heat transfer, and other related physical processes [63]. The basic procedural stages to perform a CFD simulation are shown in Figure 3.6. The pre-processing stage involves the problem identification, geometry creation, mesh generation and the specifications of the physical models, fluid properties, boundary conditions, initial conditions, solver controls etc. After the pre-processing, the solution is computed and monitored. Once the solution is converged, the results are analyzed and visualized in the form of charts, animations, vector plots, streamlines etc. in the post-processing stage. In this research work, the solution method is based on the finite volume technique in which the flow domain is divided into the control volumes and the equations are discretized and solved iteratively for each control volume [63].

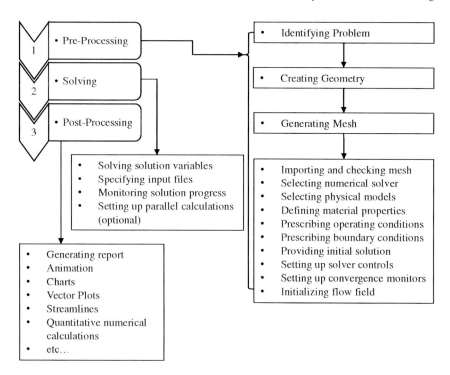

Figure 3.6:   CFD procedural stages

3.7   Codes used in this work

The codes used in the present work have been listed in Figure 3.7. Throughout this work, ANSYS DesignModeler 15.0 [59] and ANSYS Meshing 15.0 [62] have been employed for the geometry creation and mesh generation respectively. The mesh is imported into the AN-SYS CFX 15.0 [60], ANSYS FLUENT 15.0 [61] and OpenFOAM 2.3.0 [70] for setting up the respective simulation cases and solving. The converged solution data (results file) from each of the FLUENT, CFX and OpenFOAM is exported to ANSYS CFD-Post 15.0 [71] for post-processing. The ANSYS Workbench 15.0 [72] has been employed to make it easy to organize and choose the different components.

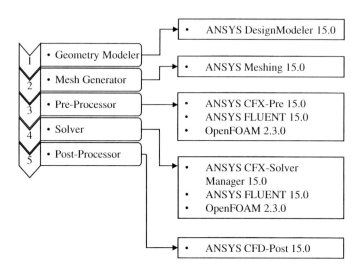

Figure 3.7:   List of codes used in this work

The brief details of the CFD programs (CFX, FLUENT and OpenFOAM) employed in this work have been summarized below.

3.7.1   CFX

ANSYS CFX 15.0 [60] is a general purpose CFD software suite that uses an element-based finite volume method and combines an advanced coupled solver (in which all the hydrodynamic equations are solved as a single system) with powerful pre- and post-processing capabilities. CFX-Pre [73], CFX-Solver [64, 65], CFX-Solver Manager [74] and CFD-Post [71] are the four software modules of ANSYS CFX. The physical models, boundary conditions, initial values and solver parameters are defined in ANSYS CFX through CFX-Pre. CFX-Solver receives the files produced by CFX-Pre and solves all the solution variables. The attributes settings for the CFD calculation, the interactive control of CFX-Solver and the information view of the emerging solution are achieved by CFX-Solver Manager. The simulation results are presented and analyzed by CFD-Post which provides state-of-the-art interactive post-processing graphics tools.

3.7.2   FLUENT

ANSYS FLUENT 15.0 [61] written in the C computer language and based on the finite volume method is a state-of-the-art CFD program ideally suited for incompressible and compressible fluid flow simulations in complex geometries [75]. All the operations including setting the boundary conditions, defining the fluid properties, executing the solution, refining the mesh, and post-processing and viewing the results are performed within the solution mode of FLUENT. However, in this work the solution data from FLUENT is exported to CFD-Post for the analysis using ANSYS Workbench.

3.7.3   OpenFOAM

OpenFOAM 2.3.0 (the Open Source Field Operation and Manipulation) [70] based on finite volume method is a C++ library, used primarily to create executables, known as applications (solvers and utilities). OpenFOAM contains a large range of solvers (e.g. for incompressible flow, compressible flow etc.) which are designed to solve a specific class of problem. The tasks involving the data manipulation are performed by the utilities (e.g. Pre-processing, mesh generation, post-processing graphics etc.). The data for mesh, fields, properties, control parameters, etc. is stored in a set of files within a case directory and cases are setup by editing the case files. Although ParaView 4.1.0 is supplied with OpenFOAM 2.3.0 for post-processing the simulation results, but in this work the solved simulation data is exported to ANSYS CFD-Post 15.0 for the easy comparison of OpenFOAM simulation results with those of CFX and FLUENT.

3.8   Concluding remarks

In this chapter, first the geometry creation and mesh generation of the base case model and the test room were described. The governing equations for the simulation of indoor air flow were then explained followed by the modeling equations for the simulation of the spread and distribution of the accidentally released pollutants into the room with the air flow. The implemented boundary conditions e.g. inlet, outlet, wall, symmetry etc. and the model parameters were then described. The basic procedural stages to perform a CFD simulation were then explained. At the end of this chapter, the description of the different CFD codes (CFX, FLUENT and OpenFOAM) used in this work was given. The next chapter will deal with the comparison of the numerical simulation results of different CFD programs for a 2D base case model of in-door air flow with and without emission source under laminar and turbulent flow conditions for the purpose of developing a better basic physical understanding and for the purpose of selection of the suitable CFD program for the further advanced simulation studies. In addition, the implementation of the developed simulation model using ANSYS CFX 15.0 will also be demonstrated.

Chapter 4   Base case simulation results

This chapter demonstrates the comparison of the numerical simulation results of different CFD programs for a 2D base case model of indoor air flow with and without emission source under laminar and turbulent flow conditions for the purpose of developing a better basic physical understanding and for the purpose of selection of the suitable CFD program for the development of the 3D transient multicomponent simulation model based on the mechanism double configuration which enables the time-dependent determination of the temporal and spatial distribution of gaseous emissions with the air flow inside the room. The first section of this chapter shows the comparison of the numerical simulation results of the three CFD programs CFX, FLUENT and OpenFOAM for the indoor air flow under both laminar flow and the turbulent flow conditions. The second section of this chapter shows the comparison of the simulation results of two CFD programs (CFX and FLUENT) to model and determine the spatial distribution of formaldehyde emission in the presence of air flow in the room under both laminar and turbulent flow conditions. This chapter concludes with the implementation of the developed simulation model based on the mechanism of double configuration using ANSYS CFX 15.0.

4.1   Indoor air flow model

Air flow is considered to be very important for the spread, transportation and distribution of the pollutants inside the room. The numerical simulations of indoor air flow under turbulent flow and laminar flow conditions are first carried out using three CFD programs (CFX, FLUENT and OpenFOAM) for the purpose of developing a basic physical understanding and for the selection of the best possible suitable and appropriate numerical simulation tool to model and simulate the complex phenomena of the spread, transportation and distribution of the pollutants inside the room in the presence of air flow. The methodology for the comparison of the three CFD programs for the indoor air flow under laminar and turbulent flow conditions is shown in Figure 4.1.

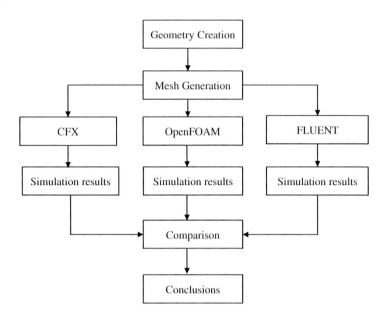

Figure 4.1:   Methodology for the comparison of the three CFD programs for the indoor air flow model under laminar and turbulent flow conditions

The specifications of the geometry dimensions have been taken from the literature [58]. The sketch of this 2D model has already been demonstrated in Figure 3.1 of Chapter 3. In order to compare the numerical simulation results of the three CFD programs, four reference lines (x = 3.0 m, x = 6.0 m, y = 0.084 m and y = 2.916 m) are defined which are shown in Figure 4.2.

As already described in Chapter 3, for the comparison of numerical simulation results of CFX, FLUENT and OpenFOAM for indoor air flow application the geometry is created in ANSYS DesignModeler 15.0, the mesh is generated in ANSYS Meshing 15.0 and the post-processing is done in ANSYS CFD-Post 15.0. The indoor air flow under laminar flow and turbulent flow conditions is solved using CFX and FLUENT, whereas in OpenFOAM the *simpleFoam* solver was used.

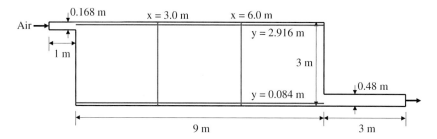

Figure 4.2: Schematic of the room [58] showing the vertical (x = 3.0 m and x = 6.0 m) and horizontal lines (y = 0.084 m and y = 2.916 m) for the measured velocity profiles

The procedure adopted in this work to use OpenFOAM is exhibited in Figure 4.3. In Open-FOAM, the data involving the mesh, fields, properties, control parameters etc. was stored in a set of files within a case directory. The case files were edited in order to setup the case in OpenFOAM. Each application was executed from the terminal command line which typically reads and writes the set of data files of the specific case. Although OpenFOAM is supplied with various mesh generation utilites e.g. *blockMesh*, *snappyHexMesh* etc., but in this work for the consistent meshing, the mesh was generated using ANSYS Meshing 15.0 and the same mesh was imported into CFX, FLUENT and OpenFOAM. The mesh data file *.msh* was written first in ASCII format and then converted into the format that OpenFOAM uses by using the mesh conversion utility *fluentMeshToFoam*.

Although OpenFOAM is supplied with a post-processing utility *paraFOAM* that employs ParaView, but in this work for the easy comparison of the numerical simulation results of the different CFD programs, ANSYS CFD-Post 15.0 has been used for post-processing the simulation results of the different CFD programs. The simulation results data from CFX and FLUENT were imported into ANSYS CFD-Post 15.0 using ANSYS Workbench 15.0. Whereas, two converters were used for the purpose of using ANSYS CFD-Post 15.0 as a post-processor for the case run in OpenFOAM. The OpenFOAM mesh was converted into FLUENT format using the conversion utility *foamMeshToFluent* and the OpenFOAM results data was converted into a *.dat* file readable by FLUENT by using *foamDataToFluent*. The OpenFOAM results data from the FLUENT was then imported into the ANSYS CFD-Post 15.0 using ANSYS Workbench 15.0. As a separate study, the parallel performance analysis of OpenFOAM running in parallel on the computing cluster for the simulation of turbulent indoor air flow is presented in Appendix B.1.3.

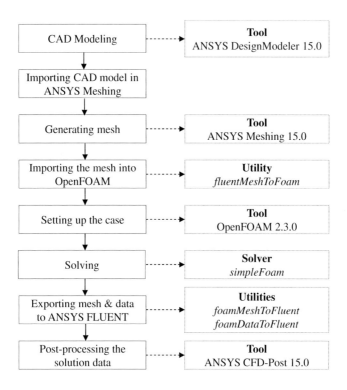

Figure 4.3:   Procedure adopted to use OpenFOAM

4.1.1   Turbulent flow

The specifications of the indoor air flow under turbulent flow condition including the measured values for the velocity distribution are taken from the literature [58]. The measured values of the velocity profiles under turbulent flow condition at the four lines (x = 3.0 m, x = 6.0 m, y = 0.084 m, y = 2.916 m) shown in Figure 4.2 are known [58] which will be compared against the simulation results for the purpose of validation of the indoor air flow simulation as will be shown in Chapter 6. The simulation settings and boundary conditions implemented separately in each of CFX, FLUENT and OpenFOAM are presented in Table 4.1. The inlet velocity perpendicular to the inlet surface is 0.455 m s^{-1}. The whole model is under atmospheric pressure, thus the relative pressure at the outlet is 0 Pa. The turbulent indoor air flow under isothermal condition (T = 298.15 K) is modelled using k-ε turbulence model. The velocity (absolute) contours from all the three CFD programs are presented in Figure 4.4. The comparison of the velocity contours from the three programs shows that the velocities predicted by these programs are almost same. The velocity is the minimum at the bottom left corner and middle of the room.

Table 4.1:   Indoor air flow (turbulent) – Simulation settings and boundary conditions implemented in CFX, FLUENT and OpenFOAM

Settings and boundary conditions	CFX	FLUENT	OpenFOAM
Simulation type	Steady state	Steady state	Steady state
Fluid information	Air	Air	Air
Solution methods	High Resolution	SIMPLE	SIMPLE
Domain	Reference Pressure: 101325 Pa Isothermal (T = 298 K) k-ε Model Scalable wall function	Operating Pressure: 101325 Pa Isothermal (T = 298 K) k-ε Model Scalable wall function	Reference Pressure: 101325 Pa Isothermal (T = 298 K) k-ε Model Standard wall function
Boundary conditions			
Velocity-Inlet	normal speed: 0.455 m s^{-1} $k = 4.97 \cdot 10^{-4}$ m^2 s^{-2} $\varepsilon = 6.59 \cdot 10^{-4}$ m^2 s^{-3}	magnitude, normal to BC: 0.455 m s^{-1} $k = 4.97 \cdot 10^{-4}$ m^2 s^{-2} $\varepsilon = 6.59 \cdot 10^{-4}$ m^2 s^{-3}	Fixed value: 0.455 m s^{-1} $k = 4.97 \cdot 10^{-4}$ m^2 s^{-2} $\varepsilon = 6.59 \cdot 10^{-4}$ m^2 s^{-3}
Pressure-Outlet	Average Static Pressure Static Pressure: 0 Pa	Gauge Pressure: 0 Pa	Fixed value: 0 Pa
Wall	No slip wall	No slip wall, Stationary Wall	No slip wall
Symmetry	Symmetry	-	Symmetry

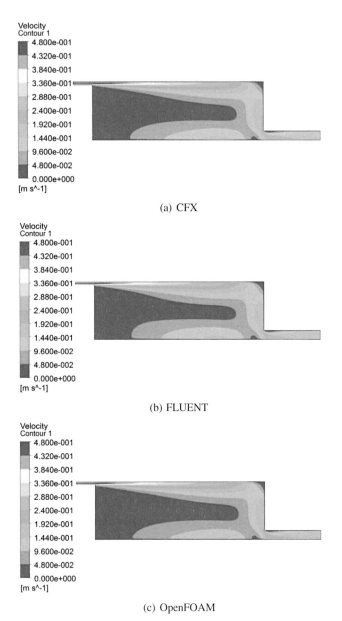

(a) CFX

(b) FLUENT

(c) OpenFOAM

Figure 4.4:   Indoor air flow (turbulent) – Comparison of the velocity contours from the
three CFD programs

The dimensionless velocities predicted by the three programs for turbulent indoor air flow at the four lines (x = 3.0 m, x = 6.0 m, y = 0.084 m and y = 2.916 m) are demonstrated in Figure 4.5. Figure 4.2 has already demonstrated the locations of the four lines (x = 3.0 m, x = 6.0 m, y = 0.084 m and y = 2.916 m) where the velocity distributions from the three CFD programs are compared. It is important to note that the velocity in the x-direction (u) is dimensioned by the inlet velocity (u_o). The simulation results of the three CFD programs for the indoor air flow under isothermal turbulent flow condition are generally in good agreement against each other as evident from Figure 4.4 and Figure 4.5. The comparison of the simulation results of the three CFD programs along the two lines at x = 3.0 m and x = 6.0 m is demonstrated in Figure 4.5 a and b respectively. The non-dimensional velocities from the three programs remain almost identical at the line x = 3.0 m, the same is true at the line x = 6.0 m. The highest velocities are predicted at y = 2.65 m and y = 0.35 m at the line x = 3.0 m as shown in Figure 4.5 a. The region at the position of y = 2.65 m denotes the maximum speed at the level of inflow and the region at y = 0.35 m the largest rate of return flow of the circulation. As presented in Figure 4.5 b, the profile of the dimensionless velocity at line x = 6.0 m becomes relatively flatter.

The comparison of the numerical simulation results of the three programs along the two lines at y = 0.084 m and y = 2.916 m is exhibited in Figure 4.5 c and d respectively. Figure 4.5 c shows that the dimensionless velocity at the line y = 0.084 m between x = 0 m to x = 8.5m is negative because of the region of the rear direction of the swirling flow, and between x = 8.5 m and 9 m the dimensionless velocity is positive due to the outflow. The deviation among the values of the non-dimensional velocities from the three CFD programs is relatively higher at the line y = 0.084 m between x = 4 m to 9 m. As demonstrated in Figure 4.5 d at the line y = 2.916 m, the deviations among the non-dimensional velocities from the three programs are very slightly significant from x = 0 m to 7 m. At the line y = 2.916 m, the velocity is higher near the inlet and decreases subsequently with the flow of the fluid in the x direction. The velocity values at all the walls are zero because of the no-slip condition at the walls. Generally, the dimensionless velocities predicted by all the three programs are in good agreement against each other.

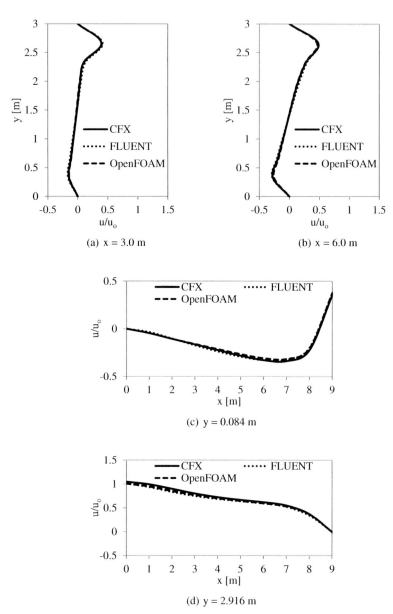

Figure 4.5:   Indoor air flow (turbulent) – Comparison of numerical simulation results (non-dimensional velocities) of the three CFD programs at the different lines locations (x = 3.0 m, x = 6.0 m, y = 0.084 m and y = 2.916 m)

4.1.2   Laminar flow

The comparison of the numerical simulation results of CFX, FLUENT and OpenFOAM for the indoor air flow under laminar flow condition is presented in Appendix B.1.1.

4.1.3   Conclusion

In this section, the numerical simulation results of three CFD programs CFX, FLUENT and OpenFOAM have been compared for the indoor air flow under both laminar flow and the turbulent flow conditions. The purpose was to develop a better understanding of indoor air flow under both turbulent and laminar flow conditions and to select the best possible suitable software package for the further relatively complex simulations. It has been seen that all these three CFD codes have almost the equal capabilities to model the indoor air flow and the simulation results of all the three programs are in good agreement against each other for both the laminar flow and the turbulent flow. Based on these results, CFX and FLUENT have been selected for the further advanced simulation studies of modeling of formaldehyde emission distribution inside the room under laminar and turbulent air flow conditions (presented in the next section) because of their availability for this research work and the ease of implementation of the complex boundary conditions and the subdomain for the formaldehyde volumetric emission source.

4.2   Indoor air flow model with an emission source

The 2D indoor air flow model described in Section 4.1 is now extended to consider a volumetric emission source placed inside the room. In addition to model and determine the spatial distribution of formaldehyde emission in the presence of air flow in the room, one of the objectives is also to see the relative capabilities of CFX and FLUENT for this multicomponent simulation. As shown in Figure 4.6, the numerical simulation results of CFX and FLUENT for this problem of formaldehyde emission inside the room in the presence of air flow will be compared, and based on these results one of these two CFD codes will be selected for the further simulations.

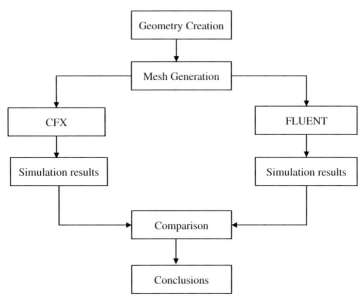

Figure 4.6:   Methodology for the comparison of the two CFD programs for the indoor air
              flow model with an emission source placed inside the room under laminar and
              turbulent flow conditions

The sketch of the model with an emission source is shown in Figure 4.7. To incorporate the volumetric source, several computational domains are setup in a model. A subdomain of 0.2 m sides is produced using ANSYS DesignModeler 15.0. The mesh of this model with an emission source is shown in Figure 4.8. After the creation of the first model, this model is frozen. Then a new sketch is created and the second volume of the body is produced. Thereafter, the created two solids are combined to form an assembly and this assembly is further used as the whole calculation area. Because the domain (chamber) and the sub-domain (source volume) were generated separately, a different element size can be adjusted for the meshing. The domain is meshed with the element size of 0.05 m and the subdomain with the element size of 0.025 m. The total number of elements is 11648.

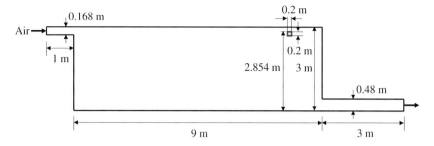

Figure 4.7:   Sketch of the 2D model with an emission source

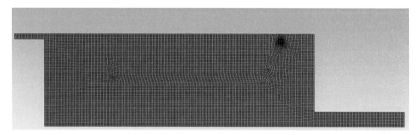

Figure 4.8:   Mesh of the 2D model with an emission source

The properties of air and formaldehyde at 338 K (65 °C) and 101325 Pa (1 atm) [76] imple-mented in both CFX and FLUENT have been tabulated in Table 4.2. For the detailed discus-sion, the work of Zheng [76] can be consulted.

Table 4.2:   Properties of air and formaldehyde at 338 K (65 °C) and 101325 Pa (1 atm) [76]

Properties	ρ	M	c_p	μ	λ	β	D_{ij}
	[kg m^{-3}]	[kg kmol^{-1}]	[J kg^{-1} K^{-1}]	[kg m^{-1} s^{-1}]	[W m^{-1} K^{-1}]	[K^{-1}]	[m^2 s^{-1}]
Air	1.0314	28.949	1008.66	$2.032 \cdot 10^{-5}$	0.02915	0.0029	$2.14 \cdot 10^{-5}$
CHOH	1.082	30.03	1244	$1.33 \cdot 10^{-5}$	0.01926	0.0029	

4.2.1 Turbulent flow

The distribution of formaldehyde emission in the presence of air flow under turbulent flow condition at steady state inside the room at 338 K (65 °C) and 101325 Pa (1 atm) is numerically determined using two CFD programs CFX and FLUENT. The value of the Reynold's number in this case is 3879. The simulation settings and boundary conditions implemented in each of CFX and FLUENT are tabulated in Table 4.3. All the boundary conditions and simulation settings are the same as presented in section 4.1.1 for the isothermal turbulent indoor air flow except an introduction of an emission source as a new source term inserted in the mass balance and also the present simulation is conducted at 338 K (65 °C). The value of the formaldehyde emission source rate is $3.475 \cdot 10^{-10}$ kg m^{-3} s^{-1} [76].

Table 4.3:   Indoor air flow (turbulent) with emission source – Simulation settings and boundary conditions implemented in CFX and FLUENT

Settings and boundary conditions	CFX	FLUENT
Simulation type	Steady state	Steady state
Fluid information	Mixture	Mixture
Solution methods	High Resolution	SIMPLE
Domain	Reference Pressure: 101325 Pa Isothermal (T = 338.15 K) k-ε Model Scalable wall function	Operating Pressure: 101325 Pa Isothermal (T = 338.15 K) k-ε Model Scalable wall function
Boundary conditions		
Velocity-Inlet	normal speed: 0.455 m s^{-1} $k = 4.97 \cdot 10^{-4}$ m^2 s^{-2} $\varepsilon = 6.59 \cdot 10^{-4}$ m^2 s^{-3}	magnitude, normal to BC: 0.455 m s^{-1} $k = 4.97 \cdot 10^{-4}$ m^2 s^{-2} $\varepsilon = 6.59 \cdot 10^{-4}$ m^2 s^{-3}
Pressure-Outlet	Average Static Pressure Static Pressure: 0 Pa	Gauge Pressure: 0 Pa
Wall	No slip wall	No slip wall, Stationary Wall
Symmetry	Symmetry	Symmetry
Emission source	CHOH: $3.475 \cdot 10^{-10}$ kg m^{-3} s^{-1}	CHOH: $3.475 \cdot 10^{-10}$ kg m^{-3} s^{-1}

The comparison of the velocity contours from the two programs for the turbulent indoor air flow with an emission source placed inside the room presented in Figure 4.9 shows that the velocities predicted by both the programs are almost same and consistent with the results presented in section 4.1.1 for the indoor air flow without any emission source placed inside the room. The current simulations are performed at the isothermal temperature of 338.15 K instead of 298.15 K. The dimensionless velocities predicted by the two programs are depicted in Figure 4.10.

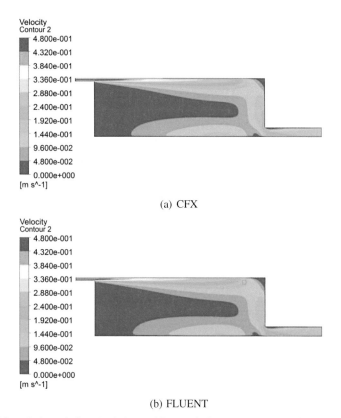

(a) CFX

(b) FLUENT

Figure 4.9:   Indoor air flow (turbulent) with an emission source – Comparison of the velocity contours from the two CFD programs

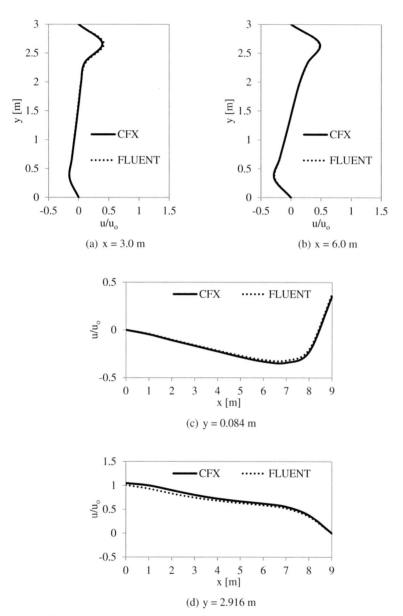

Figure 4.10: Indoor air flow (turbulent) with an emission source – Comparison of numerical simulation results (non-dimensional velocities) of the two CFD programs at the different lines locations (x = 3.0 m, x = 6.0 m, y = 0.084 m and y = 2.916 m)

The dimensionless velocities predicted by the two programs demonstrated in Figure 4.10 for turbulent indoor air flow with an emission source placed inside the room at the four lines (x = 3.0 m, x = 6.0 m, y = 0.084 m and y = 2.916 m) defined in Figure 4.2 are consistent with those of the case of only air flow inside the room described in section 4.1.1. It is to be noted that the velocity in the x-direction (u) is dimensioned by the inlet velocity (u_o). The concentration distribution of formaldehyde in terms of mass fraction in the room in the presence of air flow under turbulent condition at steady state from both the CFD programs is presented in Figure 4.11. The concentration distributions for even finer resolution range of mass fraction are demonstrated in Figure 4.12.

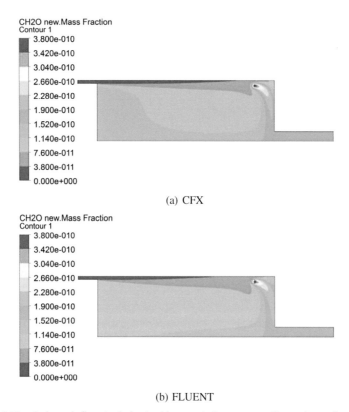

(a) CFX

(b) FLUENT

Figure 4.11:   Indoor air flow (turbulent) with an emission source – Comparison of mass fraction distributions of formaldehyde inside the room from the two CFD programs

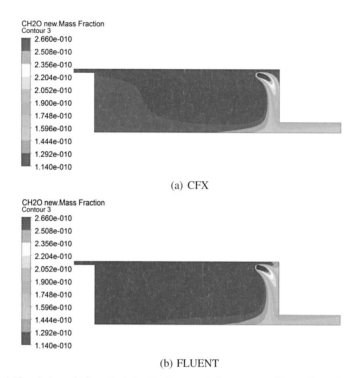

(a) CFX

(b) FLUENT

Figure 4.12:   Indoor air flow (turbulent) with an emission source – Comparison of mass
fraction distributions of formaldehyde inside the room from the two CFD
programs for the finer resolution range of mass fraction

There is no significant difference in the mass fraction distribution predicted by the two programs. Small differences in the simulation results of both programs can be found at the back
flow of the circulation and hence a slightly different distribution of formaldehyde. The difference could be caused by the fact that the loss of turbulent kinetic energy is slightly larger
when calculated with FLUENT. In general, the emissions predicted by both the programs are
transported along the fast flow to the outlet. The concentration of formaldehyde inside the
room is relatively higher near the source away from the air inlet and close to zero at the region
near to the inlet where fresh air is entering into the room. As can be seen in Figure 4.11, it is
evident that the spread of formaldehyde in the room is greatly influenced by the air flow. The
highest mass fraction of formaldehyde is $3.8 \cdot 10^{-10}$ and at the outlet it is almost between
$1.9 \cdot 10^{-10}$ and $2.28 \cdot 10^{-10}$.

4.2.2   Laminar flow

The results of the numerical simulation to determine the spatial distribution of formaldehyde emission in the presence of air flow under laminar flow condition at steady state inside the room using two CFD programs CFX and FLUENT are presented in Appendix B.1.2.

4.2.3   Conclusion

In this section, the 2D indoor air flow model was extended to introduce a volumetric emission source. In addition to model and determine the spatial distribution of formaldehyde emission in the presence of air flow under both the laminar flow condition and the turbulent flow condition in the room, one of the objectives was also to see the relative capabilities of CFX and FLUENT for the multicomponent simulation. It was found out that the air flow has a significant influence on the distribution of pollutant/s inside the room and the results from CFX and FLUENT were in good agreement with each other under both laminar flow and turbulent flow conditions. Based on these results, CFX has been selected for the further advanced simulation work because of its availability for this research work and the ease of implementation of the double configuration mechanism and the complex boundary conditions with the user defined expressions and functions.

4.3   Pollutants distribution model based on double configuration

This section explains the numerical simulation results of the developed simulation model based on double configuration using ANSYS CFX 15.0 for the prediction of the buildup of the pollutants concentration level at the ambient temperature and pressure with the passage of time at the different locations inside the room due to the accidental release of the pollutants into the room with a constant air flow and then the decrease of the pollutants concentration level inside the room with the passage of time due to air flow when the pollutants are stopped entering into the room. The two components system (propane and air) is explained followed by the three components system (propane, carbon dioxide and air). The schematic of the room along with the locations of the two data points P1 and P2 is demonstrated in Figure 4.13.

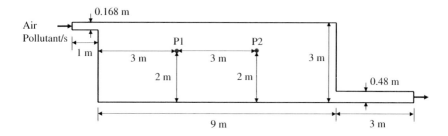

Figure 4.13:   Schematic of the room along with the locations of the two data points P1 and P2

This simulation model of the pollutants spread, transportation and distribution with the air flow inside the room under turbulent flow condition is based on the double configuration in which the steady state and transient state analyses have been sequenced in such a way that the transient simulation automatically starts upon the completion of the steady state simulation, and the results from the steady-state analysis are used as the initial guess for the transient analysis as explained in Section 3.5 of Chapter 3.

4.3.1   Two components system

The initial uniformly distributed concentration of propane present inside the room is zero whereas the propane (pollutant) is entering into the room for a certain period of time along with the air flow. The pollutant is entered into the room along with the air for the first 30 minutes. After 30 minutes, only air flow takes place. The total duration of the simulation is 40 minutes with a timestep of 1 s. The air flow rate is kept at 0.30 kg s^{-1} (turbulent flow), whereas the flowrate of the pollutant entering into the room for the first 30 minutes is 0.0003 kg s^{-1} and after 30 minutes the flow rate of the pollutant is 0 kg s^{-1}. The mass source contribution for propane has been specified by using the CFX Expression Language (CEL). The conditional "if" statement $if(t<1800[s], 0.0003[kg\ s^{-1}], 0[kg\ s^{-1}])$ has been used to specify the flow rate of propane into the room for the first 30 minutes and no flow rate of propane into the room after 30 minutes. The simulation mechanism based on the double configuration is used. All the other simulation boundary conditions, parameters and settings are the same as already ex-plained in Chapter 3. The concentration distribution of propane at the two points P1 and P2 inside the room is shown in Figure 4.14.

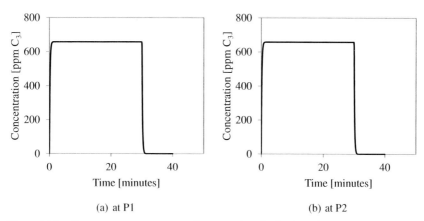

(a) at P1   (b) at P2

Figure 4.14:   Two components system – Concentration distribution of propane at P1 and P2

The concentration of propane is reached to 650 ppm in a very less time and stays constant for the remaining duration of the pollutant entering into the room. The concentration of propane suddenly falls to zero as soon as the propane inlet into the room is stopped and only air flow takes place in the room. Under ideal conditions, the efficient air flow is responsible for a sudden decrease of the propane concentration to zero. Generally speaking, the concentration distribution of propane at the two points P1 and P2 is almost the same.

The concentration distributions of propane at P1 and P2 for the periods from 0 s to 60 s and from 1800 s to 1860 s are shown in Figure 4.15 a and b respectively. The concentration of propane at P1 is less than that at P2 for a period from 0 s to 60 s because P2 is located in the way of the flow from where the fresh feed of pollutant is entering into the room with the air flow. On the other hand, for a period from 1800 s to 1860 s the concentration of propane at P1 is higher than that at P2 because P2 is located in the way of the fresh air flow which is responsible for the relatively faster removal of pollutant at P2.

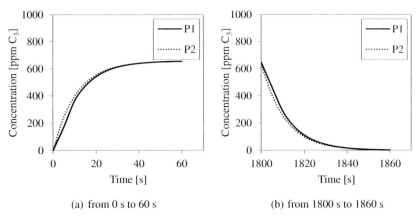

(a) from 0 s to 60 s   (b) from 1800 s to 1860 s

Figure 4.15:   Two components system – Concentration distributions of propane at P1 and P2 (a) from 0 s to 60 s and (b) from 1800 s to 1860 s

4.3.2   Three components system

The initial uniformly distributed concentrations of propane and carbon dioxide present inside the room are zero whereas both the components are entering into the room for a certain period of time along with the air flow. The pollutants are entered into the room along with the air for the first 30 minutes. After 30 minutes, only air flow takes place. The total duration of the simulation is 40 minutes with a timestep of 1 s. The air flow rate is kept at 0.30 kg s^{-1}, whereas the flowrates of propane and carbon dioxide entering into the room for the first 30 minutes are 0.0003 kg s^{-1} and 0.00015 kg s^{-1} respectively, and after 30 minutes the flow rate of both the components is 0 kg s^{-1}. The concentrations distributions of both the pollutants at the two points P1 and P2 inside the room are shown in Figure 4.16.

The mass sources contributions for propane and carbon dioxide have been specified by using the CFX Expression Language (CEL). The conditional "if" statements *if(t<1800[s], 0.0003[kg s^{-1}], 0[kg s^{-1}])* and *if(t<1800[s], 0.00015[kg s^{-1}], 0[kg s^{-1}])* for propane and carbon dioxide have been used to specify the flow rates of propane and carbon dioxide entering into the room for the first 30 minutes and no flow rate of both the components into the room after 30 minutes. The simulation mechanism based on the double configuration is used. All the other simulation boundary conditions, parameters and settings are the same as already explained in Chapter 3.

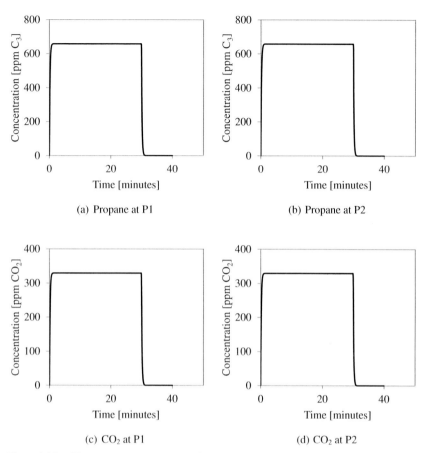

(a) Propane at P1

(b) Propane at P2

(c) CO_2 at P1

(d) CO_2 at P2

Figure 4.16:   Three components system – Concentration distributions of propane and carbon dioxide at P1 and P2

As can be seen in Figure 4.16, the concentrations of propane and carbon dioxide are reached to 650 ppm and 325 ppm respectively in a very less time and stay constant for the remaining duration of the pollutants entering into the room. The concentrations of both the components suddenly fall to zero as soon as the pollutants' inlet into the room is stopped and only air flow takes place in the room. The efficient ideal air flow is responsible for the sudden decrease of the pollutants' concentrations to zero.

4.4 Concluding remarks

In this chapter, first the numerical simulation results of three CFD programs CFX, FLUENT and OpenFOAM were compared for a simple application of indoor air flow under both laminar flow and the turbulent flow conditions. The purpose was to develop a better understanding of indoor air flow under both turbulent and laminar flow conditions and to select the best possible suitable software package for the further relatively complex simulations. It has been seen that all these three CFD codes had almost the equal capabilities to model the indoor air flow and the simulation results of all the three programs were in good agreement against each other for both the laminar flow and the turbulent flow. CFX and FLUENT were selected for the further advanced simulation studies of modeling of formaldehyde emission distribution inside the room under laminar and turbulent air flow conditions because of their availability for this research work and the ease of implementation of the subdomain for the formaldehyde volumetric emission source and the complex boundary conditions. As has been discussed in Chapter 2, adequate work has been done for the validation of the numerical simulation of the indoor air flow. In this work as demonstrated in Chapter 6, the numerical simulation of the indoor air flow under turbulent flow condition will be validated against the experimental results obtained from the literature [58].

The 2D indoor air flow model was then extended to introduce a volumetric emission source as a new source term inserted in the mass balance. In addition to model and determine the spatial distribution of formaldehyde emission in the presence of air flow under both the laminar flow condition and the turbulent flow condition in the room, one of the objectives was also to see the relative capabilities of CFX and FLUENT for the multicomponent simulation. It was found out that the air flow had a significant influence on the distribution of pollutant/s inside the room and the results from CFX and FLUENT were in good agreement with each other under both laminar flow and turbulent flow conditions. CFX was then selected for the further advanced simulation work because of its availability for this research work and the ease of implementation of the double configuration mechanism and the complex boundary conditions with the user defined expressions and functions.

After developing the better understanding of the distribution of pollutant inside the room at the different flow conditions (laminar and turbulent) of air flow and the selection of the best suitable CFD program, the numerical simulation results of the developed simulation model based on double configuration using ANSYS CFX 15.0 were presented to predict the spatial and temporal distribution of the pollutant/s entering into the room along with the air flow under turbulent flow condition. The two components system (propane and air) was explained followed by the three components system (propane, carbon dioxide and air).

The developed numerical multicomponent simulation model for the prediction of the spread, transportation and distribution of the gaseous emissions inside the room due to air flow also needs to be validated against the experimental results. For this purpose, the experimental method for both the two components system and three components system should be developed. The experiments also need to be carried out to develop a better physical understanding of pollutants spread and distribution inside the room in the presence of air flow. In the scope of this work, the experimental methodology is developed and the experiments are carried out. The next chapter explains the developed experimental setup, the measurement technique, the developed experimental procedure and the obtained experimental results. The Chapter 6 then shows the comparison of the simulation results with the experimental results for both the indoor air flow and the multicomponent flow for the purpose of validation of the developed simulation model. The application of the developed and validated simulation model will then be demonstrated in Chapter 7.

Chapter 5   Experimental investigation

This chapter describes the experimental work carried out to investigate the spread and distribution of the gaseous pollutants in the room due to their accidental release into the room with the air flow. The first section of this chapter explains the test setup. The measurement technique is described in the second section whereas the developed experimental procedure and the measurement conditions are mentioned in the third section. The results of the experiments conducted for the different investigative cases are presented and discussed in section four.

5.1   Test setup

The experiments for the investigation of the distribution of gaseous emissions with the air flow in the indoor environment were carried out in a test room developed at Fraunhofer Institute of Building Physics Stuttgart. The test chamber (Length = 4.50 m, Breadth = 4.10 m and Height = 2.57 m) shown in Figure 5.1 consists of an air inlet (1) on one side of the chamber, a flow obstruction (2) of dimensions 1.55 m × 0.60 m × 1.40 m in the middle adjacent to the wall and a suction pipe (3) on the opposite side. The air inlet and the suction pipe are exhibited in Figure 5.2. The air inlet consists of a folded spiral DN 400 through which the air stream is entered into the room by passing through a semi-circular 1.50 m high air distributor. In this distributor, the air leaves through a porous plastic fabric and flows smoothly in the room. The air is sucked through a spiral-seam DN 400 suction tube which has a length of 3.60 m and has 4 slots each 0.60 m in length. The volume flow is withdrawn through a vertically connected suction tube. As shown in Figure 5.3 a, the air guidance system on the roof of the test chamber consists of a DN 200 folded spiral containing a Wilson flow grid (1) for the volume flow measurement and a sampling probe (2) for the gas concentration measurement. The path widens at its end to a DN 400 pipe containing the duct fan (3) which maintains the air volume flow through the room. This air volume flow can be adjusted in the range of 600 to 1300 m³ h^{-1} by adjusting the speed of the duct fan. An exhaust system was installed for the safe disposal of the gases outside the building as demonstrated in Figure 5.3 b. The holes and other spacing present in the walls of the test chamber were properly covered so that the leakage of air through the walls is negligible.

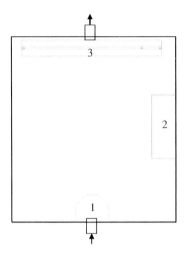

Figure 5.1:   Schematic (top view) of the test room; (1) supply air duct, (2) flow obstruction, (3) suction tube

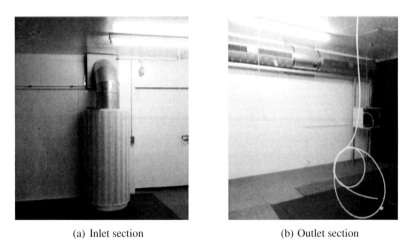

(a) Inlet section   (b) Outlet section

Figure 5.2:   Inside view of the test room; (a) Inlet section (b) Outlet section

(a) Exhaust duct system on the roof of the chamber; pressure
    measurement (1), concentration measurement (2),
    duct fan (3)

(b) Exhaust system for the safe disposal of the gases outside
    the building

Figure 5.3:   Roof of the test chamber

5.2 Measurement technique

An overview of the employed measurement technology and their respective measurement inaccuracies and measurement principles are provided in Table 5.1. The working operations and the technical details of the measurement instruments are described in Appendix A.1.

Table 5.1: An overview of the measurement technology and their respective measurement inaccuracies and measurement principles

Measurand	Manufacturer / Model / Type	Measurement principle	Measurement range	Measurement accuracy
Propane concentration	SK-Elektronik / Thermo-FID PT63 LT [77]	Flame ionization	$0.5 - 100000$ mg m^{-3} [78]	< 4 %
Carbon dioxide concentration	ABB / 3020 / Uras26 [79]	Non dispersive infrared absorption	$0 - 20$ % [80]	< 1 %
Volume flow	Halstrup Walcher / P26 Type K [81]	Differential pressure	$0 - 100$ Pa [82]	0.2 %

5.3 Experimental procedure and conditions

The schematic representations of the developed measurement procedures to investigate the concentration distribution of accidentally released gases with air into the room for two components system and three components system are shown in Figure 5.4 and Figure 5.5 respectively. The abbreviations "QIR" and "FIR" denote "Quantity Indicator and Recorder" and "Flow Indicator and Recorder" respectively.

Prior to each experiment the test chamber is always flushed with air so that the concentration level of the pollutants in the chamber becomes negligible. In each experiment, first only the air flow takes place for the first two hours to make sure that the steady state conditions are achieved followed by the gaseous pollutant inlet into the room.

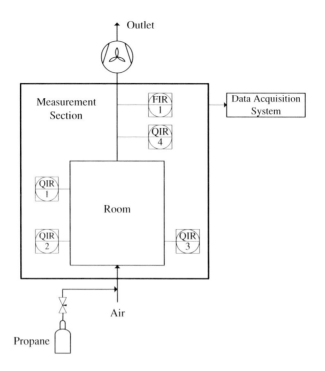

Figure 5.4:   Two components system – Schematic representation of the developed
measurement procedure to investigate the concentration distribution of the
accidentally released pollutant with air into the room

The volume flow of the air stream flowing through the room is adjusted by the speed of the
tube fan and is determined in the exhaust duct using a Wilson flow grid. Once the steady state
conditions are assumed to be achieved, the gaseous pollutant along with air is entered into the
room for a certain period of time (30 minutes). The air distributor ensures a uniform entry of
air and gaseous pollutant in the test chamber. The purpose is to investigate the spread, transportation and building up of the concentration of the gaseous pollutant at different positions in
the room with time. After 30 minutes of gaseous pollutant inlet into the room, the pollutant is
then stopped entering into the room and only the air flow takes place through the room. The
aim is to examine the decrease of the concentration of the gaseous pollutant at different locations in the room with the passage of time in the presence of the continuous air flow.

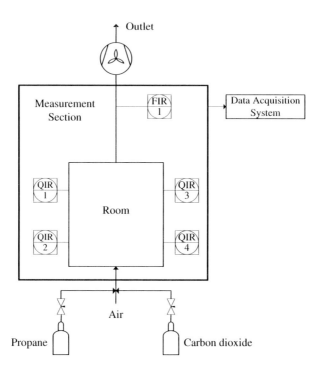

Figure 5.5:   Three components system – Schematic representation of the developed measurement procedure to investigate the concentration distribution of the accidentally released pollutants with air into the room

In one series of experiments (two components system), the concentration of the propane at different locations in the room is measured with flame ionization detectors. For data acquisition, the different programs were used to control the measurement and recording of the measurement values. Figure 5.6 shows the arrangement of the flame ionization detectors and the data acquisition system employed for this purpose. In other series of experiments (three components system), the concentrations of the propane and carbon dioxide at the different positions inside the room are measured simultaneously by flame ionization detectors and infrared photometers respectively. Figure 5.7 shows the placement of propane and CO_2 cylinders outside the room for the gases to be entered into the room.

Figure 5.6:   Flame ionization detectors (1, 2 and 3) employed to determine the propane
concentrations at different positions inside the room and the data acquisition
system (4)

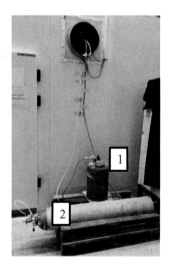

Figure 5.7:   Figure showing the placement of propane cylinder (1) and CO_2 cylinder (2)
outside the room for the gases to be entered into the room

The pressure of the propane cylinder and the initial weight of propane in the cylinder were 7.4 bar and 11 kg respectively. The pressure of the CO_2 cylinder and the initial weight of CO_2 in the cylinder were 200 bar and 45.1 kg respectively. The teflon pipes (D_i = 4 mm) were used for taking the gas sample from different locations of the room to the measurement instruments. The total duration of each experiment is different for different studied cases. At the end of each experiment, the test chamber is flushed with air and prepared again for the next experiments.

The distribution of gaseous pollutant in the test chamber in the presence of the air flow is investigated at ambient pressure and ambient temperature. The volume flow rate of the air and gaseous pollutants was adjusted at 270 L s^{-1} to 300 L s^{-1} by adjusting the speed of the exhaust fan. The valve of the propane cylinder is opened in such a way that the maximum flow of propane takes place. In a separate experimental runs (presented in detail in Appendix A.2), the weight of the propane cylinder was measured for a period of 30 minutes with an interval of 30 seconds while the propane was flowing out from the cylinder. It was noted that the weight vs. time relationship was linear. It was also noted that the flow rate of the propane releasing from the cylinder was decreasing with time, and this relationship was linear. In total, four experimental runs were conducted and it was found out that the quantity/amount of the propane released in 30 minutes for second experimental run was less than the first experimental run. The same behavior was observed for third and fourth experimental runs.

5.4 Measurement results and discussion

This section describes the experimental results of the investigation of the spread of the accidentally released gaseous pollutant/s into the room with air flow. Firstly, the results of the two components system with and without flow obstruction will be presented followed by the results of the three components system. This experimental data will also be used for the validation of the simulation model as presented in the next chapter. This section concludes with the summary of the measurement results of the further experiments considering the different case scenarios to investigate experimentally the different phenomena related to the pollutant spread in the room.

5.4.1   Two components system (without flow obstruction)

For the purpose of validation of the simulation model (two components) and the investigation of the pollutant spread without flow obstruction in the room, several experimental runs were conducted involving two components namely propane and air. The flow obstruction present in the room was removed for these experimental investigations so as to make the simulation model as simple as possible for the better comparison of the simulation results with the experimental results. Due to the limited availability of the flame ionization detectors for this research, only three FIDs could be used at a time. In order to generate enough measurement data for the simulation validation purposes, three different experimental cases involving different measurement points are discussed. The volume flow rate of the gases (air and pollutant) from the distributor is adjusted at $280 - 300$ L s^{-1}. The coordinates and locations of the measurement points for all the three cases are tabulated and shown in Table 5.2 and Figure 5.8 respectively. The measurement location "outlet" is situated at the exhaust duct system on the roof of the chamber as shown in Figure 5.3 a. It is worthy to be mentioned that for each experimental investigation studied hereinafter, the origin of the coordinates is located at the left bottom corner of the room as demonstrated in Figure 5.8. The coordinates of the locations of the measurement points for each experimental case illustrated in the next pages are taken with respect to this coordinates origin.

Table 5.2:   Two components system (without flow obstruction) – Coordinates of the measurement points for each studied case

	Measurement point 1	Measurement point 2	Measurement point 3
Case 1	(3.075, 1.000, 3.375)	outlet	(1.025, 1.000, 3.375)
Case 2	(3.075, 1.000, 2.250)	(2.050, 1.000, 2.250)	(1.025, 1.000, 2.250)
Case 3	(3.075, 1.000, 1.125)	(2.050, 1.000, 1.125)	(1.025, 1.000, 1.125)

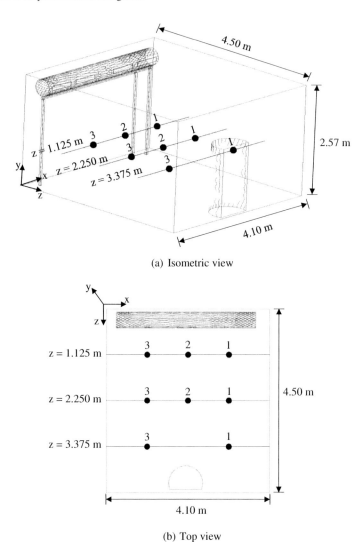

(a) Isometric view

(b) Top view

Figure 5.8: Two components system (without flow obstruction) – Locations of the measurement points for case 1 (z = 3.375 m), case 2 (z = 2.250 m) and case 3 (z = 1.125 m)

Case 1 – Measurement points near the distributor (z = 3.375 m)

In the first case of this test series, the spread of accidentally released propane with air into the room with time at ambient temperature and ambient pressure conditions is investigated. The propane concentration profiles at the two measurement locations inside the room and one measurement location at the outlet are shown in Figure 5.9. In this case, the measurement positions inside the room are placed relatively nearer to the distributor. When the steady flow conditions are assumed to be maintained with the initial air flow, the propane is fed into the room along with air for 30 minutes. After 30 minutes, the propane inlet into the room is stopped and only the air flow takes place. The total duration of this experiment is 56 minutes. The propane concentration at all the three measurement points is first increased, reached the maximum level, then decreases with time and ultimately falls to a minimum level immediately after the propane is stopped entering into the room. The concentration of propane in the room reaches to a level of around 400 ppm in the first 6 minutes. The concentration of propane at point 1 and point 3 is almost the same because the point 1 and point 3 are located inside the room at an approximately symmetrical position whereas the concentration at the outlet (position 2) is generally higher than those of point 1 and point 3 located inside the room.

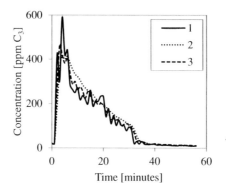

Figure 5.9:   Two components system (without flow obstruction) – Concentration distribution of propane inside the room at the measurement points located near the distributor (z = 3.375 m)

There are fluctuations in the values of the measurements because of the turbulent air flow and the non-uniform inlet flow of the propane. The propane concentration decreases with time even when the propane is entering into the room, this is because the flowrate of propane entering into the room from the cylinder is decreasing with time. At 35 minutes, the concentration of propane is 29 ppm, 26 ppm and 28 ppm at points 1, 2 and 3 respectively. At the end time of 56 minutes, the concentration at all the measurement points reduced close to 9 ppm. It takes around 26 minutes for the propane concentration to reduce from 100 ppm to 9 ppm once the propane is stopped entering into the room.

Case 2 – Measurement points in the middle of the room (z = 2.250 m)

The spread and distribution of propane into the room with time in the presence of air flow is investigated in this case when all the three measurement points are located in the middle of the room. In Figure 5.10, the propane concentration profiles at the three measurement positions at the ambient temperature and ambient pressure conditions are shown. The propane is fed into the room along with air for 30 minutes when the steady flow conditions are maintained. After 30 minutes, the propane inlet into the room is stopped and only the air flow takes place. The total duration of this experiment is 64 minutes. The propane concentration at all the three measurement points is first increased, reached the maximum level, then slightly decreases with time and ultimately falls to a minimum level immediately after the propane is stopped entering into the room. The concentration of propane in the room reaches to a level of 600 ppm in the first 5 minutes.

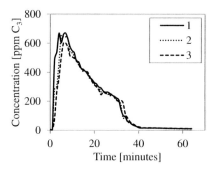

Figure 5.10:   Two components system (without flow obstruction) – Concentration distribution of propane inside the room at three measurement points located in the middle of the room (z = 2.250 m)

The concentration of propane at all the three measurement positions is generally almost the same because there is no physical hindrance present inside the room (no fireplace present inside the room). Because of the turbulent air flow and the non-uniform inlet flow of the propane, there are fluctuations in the values of the measurements. As the flowrate of propane entering into the room from the cylinder is decreasing with time, the propane concentration in the room decreases with time even when the propane is entering into the room. At 36 minutes, the concentration of propane is 55 ppm, 54 ppm and 61ppm at points 1, 2 and 3 respectively. The concentration at all the measurement points is reduced close to 10 ppm at the end time of 64 minutes.

Case 3 – Measurement points away from the distributor (z = 1.125 m)

The results of this case when the measurement points are located away from the distributor are mentioned in Appendix A.3.1.

5.4.2   Two components system (with flow obstruction)

The influence of flow obstruction present in the room on the concentration distribution of the pollutant is experimentally investigated in this section. The Table 5.3 and Figure 5.11 show the coordinates and locations respectively of the measurement points for this case. The measurement location "outlet" is situated at the exhaust duct system on the roof of the chamber as shown in Figure 5.3 a of Chapter 5. When the steady flow conditions are maintained, the propane is fed into the room along with air for 30 minutes. After 30 minutes, the propane inlet into the room is stopped and only the air flow takes place. The experiment is carried out at room temperature and pressure conditions and the total duration of this experiment is 75 minutes. The volumetric flow rate of the gases (air and pollutant) is adjusted at $270 - 280 \, \text{L s}^{-1}$.

Table 5.3:   Two components system (with flow obstruction) – Coordinates of the measurement points when the two measurement points are located away from the distributor (z = 1.125 m)

	Measurement point 1	Measurement point 2	Measurement point 3
Case 1	outlet	(1.025, 1.200, 1.125)	(3.075, 1.200, 1.125)

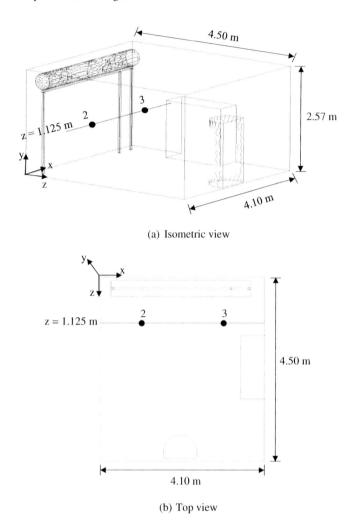

(a) Isometric view

(b) Top view

Figure 5.11: Two components system (with flow obstruction) – Locations of the measurement points when the two measurement points are located away from the distributor (z = 1.125 m)

The propane concentration profiles at the three measurement positions are shown in Figure 5.12. The concentration of propane reaches to a level of 1021 ppm, 1427 ppm and 1156 ppm at points 1, 2 and 3 in 7 minutes respectively. The concentration of propane in the room ultimately falls to a minimum level immediately after the propane is stopped entering into the room. The concentration of propane at point 2 is more than that at point 3 because the flow obstruction is present near to the point 3 which may hinder the proper flow. It is observed that the obstruction present in the way of the gas flow has a significant influence on the concentration distribution of the pollutant inside the room. At 40 minutes, the concentration of propane is 57 ppm, 74 ppm and 64 ppm at points 1, 2 and 3 respectively. Some fluctuations in the values of the measurements are observed because of the turbulent air flow and the non-uniform inlet flow of the propane. At the end time of 75 minutes, the concentration at all the measurement points reduced close to 0 ppm.

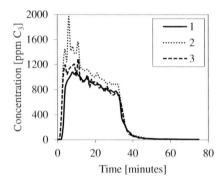

Figure 5.12:   Two components system (with flow obstruction) – Concentration distribution of propane at the three measurement points (z = 1.125 m)

5.4.3   Three components system

Several experimental runs were performed involving three components (propane, CO_2 and air) for the purpose of validation of the simulation model (three components) and the investigation of the pollutants spread in the room. In order to make the simulation model as simple as possible for the better comparison of the simulation results with the experimental results, the obstruction present in the room was removed for these experimental investigations. Two FIDs and two infrared photometers were employed for the simultaneous measurement of the concentrations of propane and CO_2 respectively at different locations inside the room. Two different experimental cases involving different measurement points are discussed in order to generate enough measurement data for the validation purpose of three components simulation. The volume flow rate of the gases from the distributor is maintained at $270 – 280$ L s^{-1}.

Case 1 – Measurement points away from the distributor (z = 2.250 m and 1.125 m)

In the first case of this test series (three components system), the simultaneous spread of propane and CO_2 with air in the room with time at ambient temperature and pressure conditions is examined when the measurement points are located away from the distributor. The coordinates and locations of the measurement points for the first case are tabulated and shown in Table 5.4 and Figure 5.13 respectively. When the steady flow conditions are maintained, the propane and CO_2 are simultaneously fed into the room along with air for 30 minutes. After 30 minutes, the propane and CO_2 inlet into the room is stopped and only the air flow takes place. The total duration of the experiment is 60 minutes.

Table 5.4:   Three components system – Coordinates of the measurement points for case 1 when the measurement points are located away from the distributor (z = 2.250 m and 1.125 m). (P = Propane, C = CO_2)

	P1 (Propane)	**P2 (Propane)**	**C1 (CO_2)**	**C2 (CO_2)**
Case 1	(1.025, 1.000, 1.125)	(1.025, 1.000, 2.250)	(3.075, 1.000, 1.125)	(3.075, 1.000, 2.250)

The propane and CO_2 concentrations profiles at the 2 FID sensors and 2 infrared photometers are shown in Figure 5.14. The propane and CO_2 concentrations at their respective measurement points within the gases inlet time are first increased, reached the maximum level, then slightly decrease with time and ultimately fall to a minimum level immediately after the gases are stopped entering into the room. The concentration of propane reaches an average level of 1200 ppm at the measurement points in 4 minutes and the maximum level of about 1550 ppm in 11 minutes as can be seen in Figure 5.14 a. The concentration of CO_2 reaches a level of around 400 ppm at the measurement points in 20 minutes as shown in Figure 5.14 b. The concentration of propane at point P2 is generally higher than the point P1 because the point P2 is located relatively nearer to the distributor from where the fresh feed is coming into the room and point P1 is located away from the distributor and at the bottom of the suction pipe. The same phenomena can be observed for the case of CO_2.

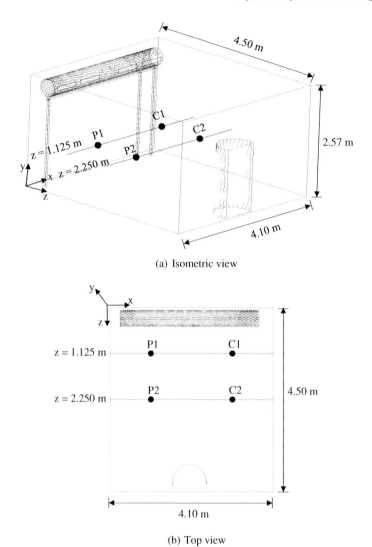

(a) Isometric view

(b) Top view

Figure 5.13:   Three components system – Locations of the measurement points for case 1
when the measurement points are located away from the distributor
(z = 2.250 m and 1.125 m).
(P = Propane, C = CO₂)

At 40 minutes, the concentration of propane is 170 ppm and 227 ppm at points P2 and P1 respectively, whereas at the same time the concentration of CO_2 is 51 ppm and 68 ppm at points
C2 and C1 respectively. There are fluctuations in the values of the measurements because of

the turbulent air flow and the non-uniform inlet flow of gases. After the gases are stopped entering into the room, the concentrations of both the gases at the positions (P2 and C2) nearer to the distributor are always slightly less than the concentrations at the positions (P1 and C1) away from the distributor. This is because the fresh air is entering into the room which removes the pollutants relatively faster at the position nearer to the distributor than at the position away from the distributor.

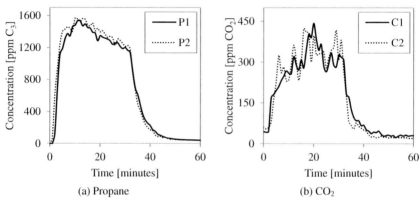

(a) Propane   (b) CO_2

Figure 5.14:   Three components system – Concentration distribution of propane and CO_2 inside the room at the measurement points located away from the distributor ($z = 2.250$ m and 1.125 m)

Case 2 – Measurement points near the distributor ($z = 3.375$ m and 2.250 m)

The simultaneous spread and distribution of propane and CO_2 in the room in the presence of air flow with time is investigated in this second case when the measurement points are located near the distributor. The coordinates and locations of the measurement points for this case is tabulated and shown in Table 5.5 and Figure 5.15 respectively.

Table 5.5:   Three components system – Coordinates of the measurement points for case 2 when the measurement points are located near the distributor
($z = 3.375$ m and 2.250 m)
(P = Propane, C = CO_2)

	P1 (Propane)	**P2 (Propane)**	**C1 (CO_2)**	**C2 (CO_2)**
Case 2	(1.025, 1.000, 3.375)	(1.025, 1.000, 2.250)	(3.075, 1.000, 3.375)	(3.075, 1.000, 2.250)

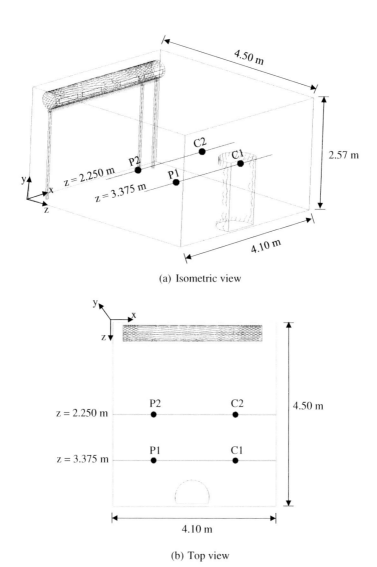

(a) Isometric view

(b) Top view

Figure 5.15: Three components system – Locations of the measurement points for case 2 when the measurement points are located near the distributor ($z = 3.375$ and 2.250 m) ($P = $ Propane, $C = CO_2$)

In Figure 5.16, the propane and CO_2 concentrations profiles at the two propane concentration measurement locations (P1 and P2) and two CO_2 concentration measurement locations (C1 and C2) respectively are shown. The propane and CO_2 are simultaneously fed into the room along with air for 31 minutes when the steady flow conditions are maintained. After 31 minutes, the propane and CO_2 inlet into the room is stopped and only the air flow takes place. The experiment is carried out at room temperature and pressure conditions and the total duration of the experiment is 52 minutes. When the gases are entering into the room, the concentrations of both the gases at their respective measurement points are first increased, reached the maximum level and then slightly decrease with time. After the gases are stopped entering into the room, the concentrations of both the gases ultimately fall to a minimum level immediately. As can be shown in Figure 5.16 a, the concentration of propane reaches to a level of 1200 ppm in 4 minutes.

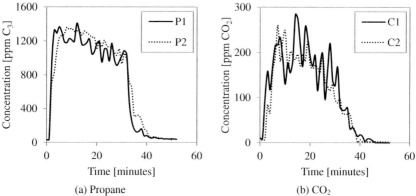

(a) Propane   (b) CO_2

Figure 5.16:   Three components system – Concentration distribution of propane and CO_2 inside the room at the measurement points located near the distributor ($z = 3.375$ m and 2.250 m)

The concentration of CO_2 reaches a level of 200 ppm in 15 minutes as exhibited in Figure 5.16 b. During the time when the gases are entering into the room, the concentration of propane at both the respective measurement positions (P1 and P2) can be said to be approximately same and in the same way the concentration of CO_2 at both the respective measurement positions (C1 and C2) are said to be approximately same, although the concentrations of the gases at their relative measurement locations nearer to the distributor are initially higher than their relative positions away from the distributor. There are fluctuations in the values of the measurements because of the turbulent air flow and the non-uniform inlet flow of the propane. At the end time of 52 minutes, the concentration at all the measurement points reduced close to 0 ppm. It can be observed that the fluctuations in the values of the measurement are relatively more at the measurement positions nearer to the distributor because of the more turbulence expected near the distributor.

5.4.4   Further experiments

The further experimental results of the spread of an accidentally released gaseous pollutant in the presence of air flow in the room described in the form of three different case scenarios to investigate the different phenomena related to the pollutant spread in the room are explained in detail in Appendix A.3. Only the important conclusions and summary of these further experimental results are mentioned here. For the case when there is no flow obstruction present inside the room, it has been revealed that the concentration of the pollutants at each relative measurement position placed inside the room was generally almost the same relative to each other because there was no physical hindrance present inside the room and the gas flow was assumed to be uniformly distributed into the room through the distributor.

The comparison of the concentrations of the pollutant at the locations nearer and away from the distributor with the passage of time when the pollutant is accidentally released into the room with air shows that the concentration of the pollutant at a point placed nearer to the distributor (from where the pollutant along with air is entering into the room) is generally higher than at the point placed relatively away from the distributor. It was also revealed that the pollutant concentration at a point placed relatively nearer to the distributor is more fluctuating than that at the point placed away from the distributor because of the more turbulent air flow nearer to the distributor. When the pollutant inlet is stopped, the propane concentration at the point relatively nearer to the distributor is slightly lower than that at the point relatively away from the distributor. This is because the air flow is higher and more turbulent at the point nearer to the distributor than the point away from the distributor hence resulting in the relatively faster rate of decrease in the concentration of propane at the point nearer to the distributor than at the point away from the distributor.

The investigation of the impact of the flow from the distributor on the concentration distribution of the pollutant at the top and bottom measurement points located at the same distance from the distributor when the pollutant is entering into the room demonstrates that the concentration of propane at the point placed at a relatively higher position is greater than that at the point placed at the relatively lower position because of the better air flow at the top than at the bottom. When the propane inlet is stopped, the propane concentration at the point placed at a relatively higher position is lower than that at the point placed at the relatively lower position because the propane is denser than air and therefore the air flow is not that efficient to remove the pollutant at the bottom with the same rate as at the top position.

5.5   Concluding remarks

The experimental setup was developed, the measurement technique was selected and several experimental runs were conducted first involving the two components (propane and air) with and without flow obstruction and then the three components (propane, CO_2 and air) for the purpose of investigation of the spread and distribution of the pollutants in the room due to the accidental release of pollutants into the room with the air flow and for the experimental validation of the developed simulation model (two components and three components) based on double configuration.

The flame ionization detectors and infrared photometers were employed to determine the concentrations of propane and CO_2 respectively at the different locations in the room. It has been revealed that the concentration of the pollutants at each relative measurement position placed inside the room for the cases when there was no flow obstruction present inside the room was generally almost the same relative to each other because there was no physical hindrance present inside the room and the gas flow was assumed to be uniformly distributed into the room through the distributor. The concentration of the pollutants at all the three measurement points was first increased, reached the maximum level, then slightly decreased with time and ultimately fallen to a minimum level immediately after the pollutants were stopped entering into the room.

The experimental investigation of the influence of the flow obstruction present in the room on the concentration distribution of the pollutant showed that the obstruction present in the way of the gas flow had a significant influence on the concentration profiles of the pollutant inside the room.

Extensive experimental data has been generated for the experimental validation of the developed simulation model. The next chapter deals with the comparison of the experimental results with the simulation results for the purpose of experimental validation of the developed simulation model (for both two components and three components systems).

Chapter 6   Comparison of simulation results with experimental results

In this chapter, the comparison of the numerical simulation results of the three CFD programs with the existing experimental data to determine the velocity distribution inside the room is demonstrated in the first section. The comparison of the simulation results with the experimental results of both the two components system (with and without flow obstruction) and the three components system for the purpose of experimental validation of the developed simulation model based on double configuration is then presented followed by the summary and conclusion of this chapter.

6.1   Indoor air flow model

In this section, the simulation results of the three CFD programs to model the indoor air flow presented in Chapter 4 are compared with the existing measurement results taken from the literature [58]. It is to be noted that the simulation results from ANSYS CFX 11.0 have been compared against the measurement values by Rong et al. [83]. The comparison of the measurement results with the simulation results from Star-CCM+ has been done by Dréau et al. [84]. In the present work, the comparison of the simulation results of the three CFD codes with the measurement values is done. The comparison of the simulation results of CFX, FLUENT and OpenFOAM with the measurement results for the non-dimensional velocity along the four lines (x = 3.0 m, x = 6.0 m, y = 0.084 m and y = 2.916 m) exhibited in Figure 4.2 is shown in Figure 6.1. It is important to note that the velocity in the x-direction (u) is dimensioned by the inlet velocity (u_o). It is observed that the simulation results of the three CFD programs are generally in good agreement with the measurements. Figure 6.1 a and b shows the comparison of the simulation results with the measurement results along the two lines at x = 3.0 m and x = 6.0 m respectively. The simulation results are in better agreement at x = 3.0 m as compared to those at x = 6.0 m. The programs slightly overestimate the non-dimensional velocity at the middle region of line x = 6.0 m.

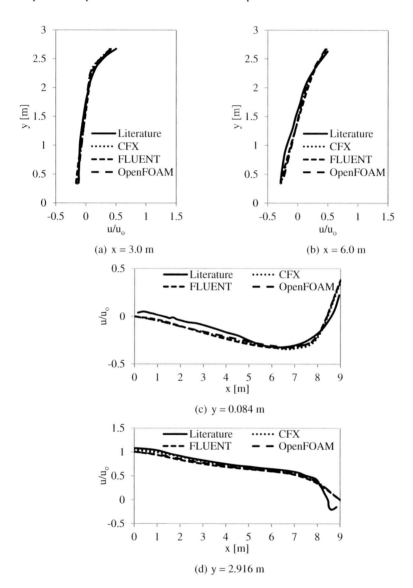

(a) x = 3.0 m

(b) x = 6.0 m

(c) y = 0.084 m

(d) y = 2.916 m

Figure 6.1:   Comparison of simulation results (dimensionless velocities) of three CFD codes with the measurement [58] at the four different lines locations (x = 3.0 m, x = 6.0 m, y = 0.084 m and y = 2.916 m)

The comparison of the simulation results with the measurement results along the two lines at y = 0.084 m and y = 2.916 m is demonstrated in Figure 6.1 c and d respectively. The three CFD programs slightly underpredict the velocity at the lower left corner of the room. The programs do not predict accurately the non-dimensional velocity at the upper right corner of the room. The programs overpredict the velocity at the upper right corner of the room. All these results are consistent with the results reported in the literature [83, 84].

6.2 Developed simulation model based on double configuration

In this section, the simulation results of the two components system (with and without flow obstruction) and three components system are compared with the corresponding experimental results for the purpose of validation of the developed simulation model. The simulated concentration distributions are compared against the experimentally obtained concentration distributions. The experimental results have already been presented in Chapter 5. The developed simulation model based on the double configuration already explained in Chapter 3 and implemented in Section 4.3 of Chapter 4 is employed to compare the simulation results with the experimental results for both the two components system and the three components system.

6.2.1 Two components system (without flow obstruction)

For the purpose of validation of the two components simulation model (without flow obstruction), the simulation results of the three different cases based on the different concentration measurement points inside the room are compared with the corresponding experimental results presented in Chapter 5. The locations of the measurement points have already been exhibited in Figure 5.8 of Chapter 5. The inlet mass flow rate of air for all the three cases is 0.34 kg s^{-1}. The mass source contribution for propane for all the three cases has been specified by using the CFX Expression Language. The time dependent expressions have been formulated for the boundary mass source contribution of propane for all the cases. An example for the selection of the time dependent expression to specify the propane mass source contribution (flow rate of the propane) for case 1 when the measurement points are located near the distributor (z = 3.375 m) is explained in Appendix B.2.1. The same methodology as demonstrated in Appendix B.2.1 was adopted to determine the best expression for the specification of the component/s boundary source/s for all the other cases of two components simulation and three components simulation. For the two components system, the obtained expressions and the conditional statements of propane boundary sources for each of the three cases are mentioned in Table 6.1. As can be seen in Table 6.1, the conditional "if" statement is used to specify the flow rate of propane into the room for the first 30 minutes and no flow rate of propane into the room after 30 minutes. The total time duration of the simulations for case 1, case 2 and case 3 are 3360 s, 3840 s and 2880 s respectively.

Table 6.1:   Two components system (without flow obstruction) – Boundary mass source contribution of propane for all the three cases

	Conditional statement	**expression**
Case 1	if(t<1800[s], **expression**, 0[kg s^{-1}])	-(0.000000094[kg s^{-2}]*t)+0.000204[kg s^{-1}]
Case 2	if(t<1800[s], **expression**, 0[kg s^{-1}])	-(0.00000018[kg s^{-2}]*t)+0.0003774[kg s^{-1}]
Case 3	if(t<1800[s], **expression**, 0[kg s^{-1}])	-(0.00000036[kg s^{-2}]*t)+0.0004182[kg s^{-1}]

The influence of diffusion on the concentration distribution of the pollutant in the room when the pollutant is entering into the room with the turbulent air flow is demonstrated in Appendix B.2.2. The influence of gravity on the temporal and spatial distribution of pollutant inside the room with the air flow was studied and has not been included in this written dissertation. It revealed that the gravity had a significant influence on the distribution of pollutant inside the room with the air flow. For this reason, the influence of gravity has been implemented in the simulation model using the buoyancy approach [64, 65]. For this purpose, a source term $S_{M,buoy}$ given by Eq. (3.19) was added to the momentum equations. The different sensitivity analyses were conducted to select the best combinations of residual target, turbulent intensities and timesteps. All the simulations of the 3D test room presented in this thesis are conducted with the residual target of 10^{-4}, timesteps of 1 s and a turbulent intensity of 5 %. In CFX, the advection scheme selected is High Resolution and the transient scheme employed is Second Order Backward Euler. The evaluation of different turbulence models (standard k-ε, RNG k-ε, k-ε EARSM and BSL) on the simulation results was also done and has not been given in this written thesis. This evaluation study revealed that the influence of the selection of the turbulence models on the simulation results of our particular problem was not significant; hence the k-ε turbulence model being the standard model of practical technical flow calculations has been selected to be implemented in all the simulations shown in this work.

Case 1 – Measurement points near the distributor (z = 3.375 m)

The simulation results of two components system are compared with the experimental results for the case when the measurement points are located relatively nearer to the distributor. The measured and simulated concentration profiles of propane inside the room at the two measurement locations plotted for a period of 3360 s in Figure 6.2 generally shows that the measured and simulated concentrations of propane are first increased, reached the maximum level, then decreased with time and ultimately fallen to a minimum level immediately after the propane is stopped entering into the room. The simulation and experimental results generally show the same response with time however the simulated propane concentration reaches to the maximum level slightly faster at the start of the propane inlet into the room because, contrary to the reality, the simulation is carried out at the ideal conditions in which the pollutants are efficiently distributed into the room at a faster rate by the efficient air flow once the pollutant starts entering into the room.

The percentage difference between the simulated and experimental concentrations ($\%_{\text{diff}}$) is given by the following equation;

$$\%_{\text{diff}} = \left(\frac{C_{\text{sim}} - C_{\text{exp}}}{C_{\text{exp}}}\right) \times 100 \qquad (6.1)$$

The comparison of the propane concentration profiles at the measurement point 1 is exhibited in Figure 6.2 a. The difference between simulation and experimental concentrations at point 1 at 18 minutes and 24 minutes are 8.6 ppm and 9.6 ppm respectively. The average percentage difference between simulated and experimental concentrations at point 1 for the period when the propane is entering into the room is 16 %. In Figure 6.2 b, the propane concentration profiles at the measurement point 3 are compared. The diagram shows that the difference between simulation and experimental concentrations at point 3 at 18 minutes and 24 minutes are 5 ppm and 4 ppm respectively. For the period when the propane is entering into the room, the average percentage difference between simulated and experimental concentrations at point 3 is 9.6 %. The simulated propane concentration at both the points 1 and 3 immediately reaches to the zero concentration level relatively fast after the propane inlet into the room is stopped because, contrary to the reality, the simulation is carried out at the ideal conditions in which the pollutants are efficiently removed out of the room by the efficient air flow once the pollutant is stopped entering into the room. The simulated propane concentration is not fluctuating because the pollutant is efficiently distributed uniformly into the room by the simulated ideal air flow.

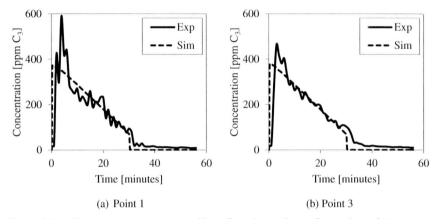

(a) Point 1   (b) Point 3

Figure 6.2:   Two components system (without flow obstruction) – Comparison of the
simulation results with the measurement results for case 1 when the
measurement points are located near the distributor (z = 3.375 m)

Case 2 – Measurement points in the middle of the room (z = 2.250 m)

The measured and simulated concentration profiles of propane inside the room for a period of 3840 s are plotted in Figure 6.3 for the case when the measurement points are located in the middle of the room. The simulation and experimental results generally show the same response with time. The comparison of the propane concentration profiles exhibited in Figure 6.3 a at the measurement point 1 shows that the difference between simulation and experimental concentrations at point 1 at 20 minutes and 24 minutes are 11 ppm and 15 ppm respectively, whereas the average percentage difference between simulated and experimental concentrations at point 1 between 4 minutes and 26 minutes when the propane is entering into the room is 14.7 %. The propane concentration profiles at the measurement point 2 compared in Figure 6.3 b shows that the difference between simulation and experimental concentrations at point 2 at 20 minutes and 24 minutes are 39 ppm and 0.46 ppm respectively. For the period when the propane is entering into the room, the average percentage difference between simulated and experimental concentrations at point 2 between 4 minutes and 26 minutes is 9.3 %.

The propane concentration profiles at the measurement point 3 compared in Figure 6.3 c shows that the difference between simulation and experimental concentrations at point 3 at 20 minutes and 24 minutes are 13.8 ppm and 20.7 ppm respectively. For the period when the propane is entering into the room, the average percentage difference between simulated and experimental concentrations at point 3 between 4 minutes and 26 minutes is 9 %.

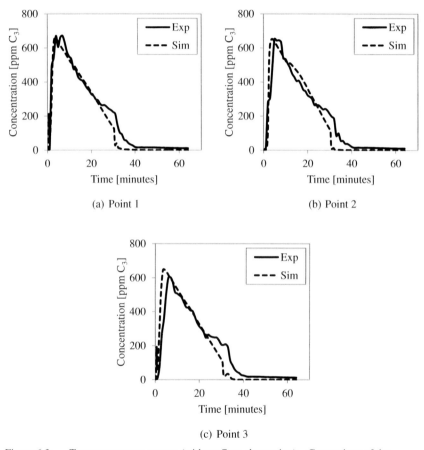

(a) Point 1   (b) Point 2

(c) Point 3

Figure 6.3:   Two components system (without flow obstruction) – Comparison of the
               simulation results with the measurement results for case 2 when the
               measurement points are located in the middle of the room ($z = 2.250$ m)

Case 3 – Measurement points away from the distributor (z = 1.125 m)

The comparison of the simulation results with the measurement results for this case when the concentration measurement points are located away from the distributor is demonstrated in Appendix B.2.3.

6.2.2 Two components system (with flow obstruction)

In this section, the simulation results of the two components system are compared with the corresponding experimental results for the purpose of validation of the developed simulation model when the flow obstruction is present in the room. The experimental results for this case have already been presented in Section 5.4.2. The coordinates and locations of the measurement points have already been exhibited in Table 5.3 and Figure 5.11 respectively of Section 5.4.2. The inlet mass flow rate of air for this case is 0.31 kg s^{-1}. Table 6.2 shows the mass source contribution for propane for this case specified by using the same methodology as demonstrated in Section 6.2.1. The total time duration of the simulation for this case is 75 minutes. All the other boundary conditions, simulation parameters and settings are the same as mentioned in Section 6.2.1.

Table 6.2: Two components system (with flow obstruction) – Boundary mass source contribution of propane

Conditional statement	expression
if(t<1800[s], **expression**, 0[kg s^{-1}])	-(0.00000016[kg s^{-2}]*t)+0.00066[kg s^{-1}]

The comparison of the propane concentration profiles exhibited in Figure 6.4 a for this case shows that the difference between simulation and experimental concentrations at point 2 at 16 minutes and 26 minutes are 6.36 ppm and 26.22 ppm respectively. For the period when the propane is entering into the room, the average percentage difference between simulated and experimental concentrations at point 2 between 3 minutes and 30 minutes is 6.25 %. It is pertinent to note that the simulation and experimental results generally show the same response with time. Figure 6.4 b shows that the difference between simulation and experimental concentrations at point 3 at 16 minutes and 26 minutes are 74.12 ppm and 43.87 ppm respectively. The average percentage difference between simulated and experimental concentrations at point 3 between 3 minutes and 30 minutes is 15.04 % for the period when the propane is entering into the room.

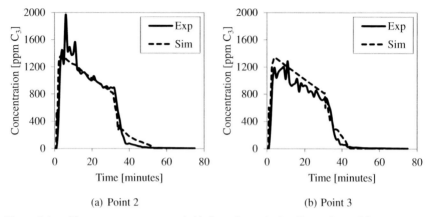

(a) Point 2   (b) Point 3

Figure 6.4:   Two components system (with flow obstruction) – Comparison of the
simulation results with the measurement results when the two measurement
points are located away from the distributor (z = 1.125 m)

6.2.3   Three components system

In this series of three components system, the simulation results of the two different cases based on the different concentration measurement points inside the room are compared with the corresponding experimental results presented in Section 5.4.3 for the purpose of validation of the three components simulation model. The locations of the concentrations measurement points of case 1 and case 2 have already been exhibited in Figure 5.13 and Figure 5.15 of Chapter 5 respectively. The inlet mass flow rate of air for all the two cases is 0.318 kg s^{-1}. The mass source contributions for each of propane and carbon dioxide for all the two cases has been specified by using the same method as presented in the two components system case. The total time duration of the simulations for case 1 and case 2 are 3600 s and 3120 s respectively. The expressions and conditional statements for each of propane and carbon dioxide boundary sources for the two cases are shown in Table 6.3 and Table 6.4 respectively.

Table 6.3:   Three components system – Boundary mass source contribution for propane

	Boundary source (propane)	expression
Case 1	if(t<1800[s], expression, 0[kg s^{-1}])	-(0.000000086[kg s^{-2}]*t)+0.0007905[kg s^{-1}]
Case 2	if(t<1860[s], expression, 0[kg s^{-1}])	-(0.000000084[kg s^{-2}]*t)+0.0006885[kg s^{-1}]

Table 6.4:   Three components system – Boundary mass source contribution for carbon dioxide

	Boundary source (CO_2)	expression
Case 1	if(t<1800[s], expression, 0[kg s^{-1}])	-(0.0000000 [kg s^{-2}]*t)+0.000154[kg s^{-1}]
Case 2	if(t<1860[s], expression, 0[kg s^{-1}])	-(0.000000010[kg s^{-2}]*t)+0.000099[kg s^{-1}]

Case 1 – Measurement points away from the distributor (z = 2.250 m and 1.125 m)

For the case when the concentration measurement points are located away from the distributor, the comparison of the propane concentration profiles exhibited in Figure 6.5 a at the measurement point P1 shows that the difference between simulation and experimental concentrations at 12 minutes and 40 minutes are 9.7 ppm and 4.2 ppm respectively. The average percentage difference between simulated and experimental concentrations at point P1 between 5 minutes and 30 minutes when the propane is entering into the room is 9.16 %. The difference between simulation and experimental concentrations at point P2 at 12 minutes and 40 minutes are 12.72 ppm and 33.3 ppm respectively as demonstrated in Figure 6.5 b. For the period when the propane is entering into the room, the average percentage difference between simulated and experimental concentrations at point P2 between 5 minutes and 30 minutes is 4.63 %.

The comparison of the carbon dioxide concentration profiles demonstrated in Figure 6.5 c at the measurement point C1 shows that the difference between simulation and experimental concentrations at point C1 at 15 minutes and 30 minutes are 8 ppm and 6.58 ppm respectively. The average percentage difference between simulated and experimental concentrations at point C1 between 5 minutes and 30 minutes when the pollutants are entering into the room is 18.3 %. The carbon dioxide concentration profiles at the measurement point C2 are compared in Figure 6.5 d. The diagram shows that the difference between simulation and experimental concentrations at point C2 at 15 minutes and 30 minutes are 17.73 ppm and 10.17 ppm respectively. For the period when the pollutants are entering into the room, the average percentage difference between simulated and experimental concentrations at point C2 between 5 minutes and 30 minutes is 14.96 %. The simulated carbon dioxide concentration is not fluctuating because the pollutants are efficiently distributed uniformly into the room by the simulated ideal air flow. The simulated carbon dioxide concentration at all the two measurement points immediately reaches to the zero concentration level relatively fast after the pollutants inlet into the room is stopped because, contrary to the reality, the simulation is carried out at the ideal conditions in which the pollutants are efficiently removed out of the room by the efficient air flow once the pollutant is stopped entering into the room.

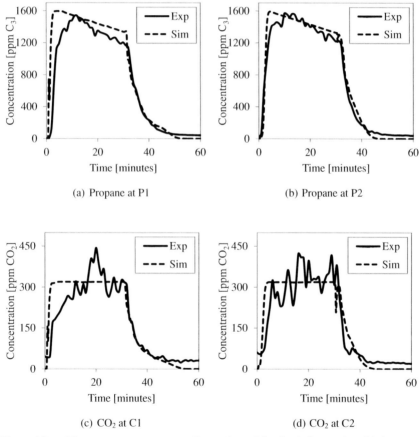

(a) Propane at P1

(b) Propane at P2

(c) CO_2 at C1

(d) CO_2 at C2

Figure 6.5:   Three components system – Comparison of the simulation results with the measurement results for case 1 ($z = 2.250$ m and 1.125 m)

Case 2 – Measurement points near the distributor (z = 3.375 m and 2.250 m)

The comparison of the propane concentration profiles exhibited in Figure 6.6 a for the case when the measurement points are located near the distributor shows that the difference between simulation and experimental concentrations at point P1 at 13 minutes and 26 minutes are 13.97 ppm and 17.58 ppm respectively. For the period when the propane is entering into the room, the average percentage difference between simulated and experimental concentrations at point P1 between 5 minutes and 30 minutes is 10.07 %.

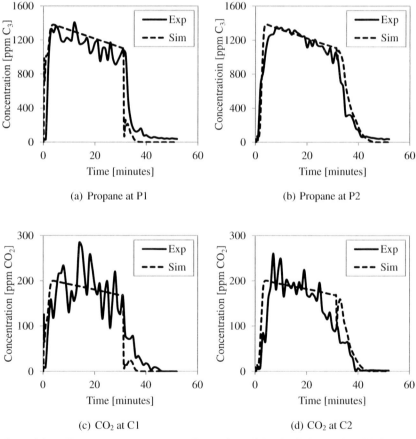

(a) Propane at P1   (b) Propane at P2

(c) CO_2 at C1   (d) CO_2 at C2

Figure 6.6:   Three components system – Comparison of the simulation results with the measurement results for case 2 (z = 3.375 m and 2.250 m)

Figure 6.6 b shows that the difference between simulation and experimental concentrations at point P2 at 13 minutes and 26 minutes are 35.02 ppm and 35.85 ppm respectively. The average percentage difference between simulated and experimental concentrations at point P2 between 5 minutes and 30 minutes is 4.27 % for the period when the propane is entering into the room. The difference between simulation and experimental concentrations of carbon dioxide at point C1 at 13 minutes and 24 minutes are 6.64 and 0.35 ppm respectively as shown in Figure 6.6 c. The average percentage difference between simulated and experimental concentrations at point C1 between 5 minutes and 30 minutes when the pollutants are entering into the room is 20 %. Figure 6.6 d shows that the difference between simulation and experimental concentrations at point C2 at 13 minutes and 24 minutes are 3.64 ppm and 20.50 ppm respectively. For the period when the pollutants are entering into the room, the average percentage difference between simulated and experimental concentrations at point C2 between 5 minutes and 30 minutes is 17 %.

6.3   Concluding remarks

The simulation results of CFX, FLUENT and OpenFOAM for the non-dimensional velocities along the four lines (x = 3.0 m, x = 6.0 m, y = 0.084 m and y = 2.916 m) in the room were in good agreement with the measurement results taken from the literature [58]. Based on this comparison, it can be concluded that all the three employed CFD programs give relatively accurate results and have the equal capabilities to model the indoor air flow.

After the validation of the simulation model of indoor air flow, the developed numerical simulation model based on double configuration for the determination of the gaseous pollutant spread and transportation into the room with the air flow was validated against the experimental data. The simulation and experimental results were in good agreement. As has been seen, the difference between simulation and experimental results was not more than 20 percent which is an acceptable standard for the CFD validations. The simulation and experimental results generally showed the same response with time, however the simulated propane concentration at all the measurement points immediately reached to the zero concentration level relatively fast after the propane inlet into the room was stopped because, contrary to the reality, the simulation was carried out at the ideal conditions in which the pollutants were efficiently removed out of the room by the efficient air flow once the pollutant was stopped entering into the room. The simulated propane concentration was not fluctuating because the pollutant was efficiently distributed uniformly into the room by the simulated ideal air flow.

As the developed simulation model has been validated in this chapter, the next chapter will deal with the application of the developed simulation model to the loss prevention and risk mitigation in the indoor environment and to study the influence of different parameters on the pollutant spread and distribution inside the room in the presence of air flow which would otherwise be very difficult to be studied with the help of the expensive and time consuming experiments.

Chapter 7   Further simulation studies

The further simulation studies have been carried out to apply the developed simulation model to the loss prevention and risk mitigation in the indoor environment and to study the influence of different parameters on the pollutant spread and distribution inside the room with the air flow which would otherwise be very difficult to be studied with the help of the expensive and time consuming experiments. In this chapter, various cases have been studied with the same locations of the data points inside the room. The coordinates and locations of the data points of all the studied cases relative to the origin of the coordinates are tabulated and shown in Table 7.1 and Figure 7.1 respectively. The distributor is located relatively nearer to the data point 1 than that of the data point 2.

Table 7.1:   Further simulation studies – Coordinates of the data points for all the studied cases

	Data point 1	Data point 2
Simulation cases	(2.050, 1.000, 2.250)	(2.050, 1.000, 1.125)

7.1   Case 1 – Influence of air change rate

The influence of different mass flow rates of air entering into the room through the distributor at the ambient temperature and pressure conditions on the rate of decrease of propane concentration in the room is studied when the propane is initially distributed uniformly into the room with a certain concentration. The Lower Explosive Limit (LEL) of propane is 21000 ppm [85]. The objective of this study is to analyze the influence of different air change rates on the rate of concentration decrease of propane when the propane in the room is accidentally reached to a concentration level close to its LEL value.

93

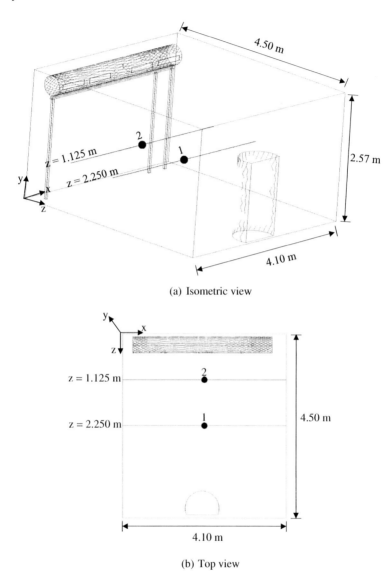

(a) Isometric view

(b) Top view

Figure 7.1: Further simulation studies – Locations of the data points for all the studied cases

The recommended air change rate in the factory buildings, fumes and moisture is $10 - 15$ hr^{-1} whereas in the kitchens is $15 - 60$ hr^{-1} [86]. The air change rates (ACH) with the corresponding mass flow rates selected for this study are mentioned in Table 7.2. The total duration of the simulations is 3600 s (60 mins) with a timestep of 1 s. The transient simulations are carried out only, and the simulation mechanism based on the double configuration explained in Chapter 3 and implemented in Chapter 4 and Chapter 6 is not used. All the other simulation boundary conditions, parameters and settings are the same as already explained in Chapter 3 and implemented in Chapter 4 and Chapter 6.

Table 7.2:   Air change rates with the corresponding mass flow rates

Mass flow rate (kg s^{-1})	Air Change Rate (ACH)
0.155	10
0.310	20
0.460	30
0.610	40
0.770	50
0.930	60

Figure 7.2 shows the relative concentration decrease with time at the different locations inside the room due to the different air change rates when the initial concentration of propane inside the room is 20000 ppm. The concentration of propane generally decreases from a level of 20000 ppm to about 0 ppm with the passage of time. The rate of decrease of propane concentration in the room is fastest with 60 ACH and the slowest with the 10 ACH. The concentration of propane at point 2 is generally higher than that of point 1 because fresh air is entering into the room closer to point 1 as compared to point 2. In the first 4 minutes, the concentration of propane inside the room at point 1 is reached to 14690 ppm for 10 ACH, 7858 ppm for 20 ACH, 6448 ppm for 30 ACH, 5784 ppm for 40 ACH, 4147 ppm for 50 ACH and 1236 ppm for 60 ACH as shown in Figure 7.2 a. Within 10 minutes, the concentration of propane at point 1 is reduced to approximately 0 ppm for 60 ACH and 6740 ppm for 10 ACH. The concentration of propane at point 1 becomes zero in 37 minutes for 20 ACH, in 24 minutes for 30 ACH, in 17 minutes for 40 ACH and in 12 minutes for 50 ACH. The concentration of propane is still at 220 ppm for 10 ACH at point 1 in 60 minutes.

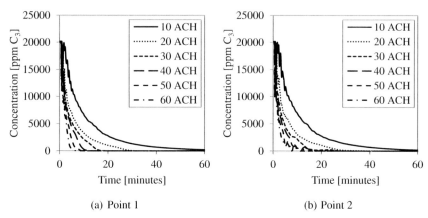

(a) Point 1   (b) Point 2

Figure 7.2: Case 1 "Influence of air change rate" – Concentration distribution of propane inside the room when the initial concentration of propane inside the room is 20000 ppm

The NIOSH Immediately Dangerous to Life or Health (IDLH) concentration of propane is 2100 ppm [87]. The purpose of this next study is to predict the influence of different air change rates on the relative concentration decrease with time at the different locations inside the room when the initial concentration of propane inside the room is accidentally reached to its IDLH value of 2100 ppm. The concentration distributions of propane with time at the two different locations inside the room with the different air change rates are demonstrated in Figure 7.3. Within 2 minutes the concentrations of propane are reached to 2098 ppm and 103 ppm for 10 ACH and 60 ACH respectively as shown in Figure 7.3 a. In 5 minutes the concentration becomes 1211 ppm and 6.5 ppm for 10 ACH and 60 ACH respectively. At 40 minutes, the concentration is reached to 57 ppm at point 1 and 74 ppm at point 2 for 10 ACH. The concentration of propane at point 2 is generally higher than that of point 1 because fresh air is entering into the room closer to point 1 as compared to point 2. These results show that the air change rate has a significant influence on the rate of decrease of propane concentration inside the room.

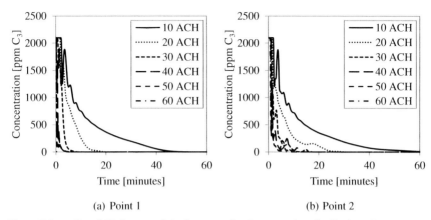

(a) Point 1   (b) Point 2

Figure 7.3:   Case 1 "Influence of air change rate" – Concentration distribution of propane
               inside the room when the initial concentration of propane inside the room is
               2100 ppm

Figure 7.4 shows the concentration distribution of propane at two different locations inside
the room when the initial uniformly distributed concentration of propane inside the room is
500 ppm. The four different flow rates of air are 0.26 kg s^{-1}, 0.30 kg s^{-1}, 0.34 kg s^{-1} and
0.40 kg s^{-1}. The total duration of the simulations is 900 s with a timestep of 1 s.

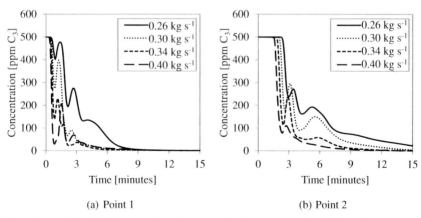

(a) Point 1   (b) Point 2

Figure 7.4:   Case 1 "Influence of air change rate" – Concentration distribution of propane
               inside the room when the initial concentration of propane inside the room is
               500 ppm

As can be seen in Figure 7.4, the rate of decrease of propane concentration in the room is fastest with 0.40 kg s^{-1} and the slowest with the 0.26 kg s^{-1}. The concentration of propane at point 2 is generally higher than that of point 1 because fresh air is entering into the room closer to point 1 as compared to point 2. The concentration of propane at all the flowrates is reduced approximately close to zero level in 9 minutes at point 1 as shown in Figure 7.4 a. Whereas, at point 2 the concentration of propane is even not reduced to zero level for the flow rate of 0.26 kg s^{-1} in 15 minutes as exhibited in Figure 7.4 b.

7.2   Case 2 – Influence of different initial in-room pollutant concentrations

The influence of different initial uniformly distributed concentrations of propane inside the room on the relative rates of decrease of propane concentrations in the room at the ambient temperature and pressure conditions with a constant mass flow rate of air entering into the room through the distributor is studied. The purpose of this study is to predict the relative concentration decrease with time at the different locations inside the room due to the different initial propane concentrations present inside the room at a constant air flow rate. The four different initial concentrations of propane present inside the room are 100 ppm, 300 ppm, 500 ppm and 700 ppm, whereas the air flow rate through the distributor is kept constant at 0.34 kg s^{-1}. The total duration of the simulations is 900 s with a timestep of 1 s. The transient simulations are carried out only, and the simulation mechanism based on the double configuration explained in Chapter 3 and implemented in Chapter 4 and Chapter 6 is not used. All the other simulation boundary conditions, parameters and settings are the same as already explained in Chapter 3 and implemented in Chapter 6.

The concentration profiles of propane at the two locations inside the room are presented in Figure 7.5. The concentration of propane generally decreases from its initial concentration to the lowest concentration with the passage of time due to the air flow. As shown in Figure 7.5 a, the concentration of propane at point 1 approaches approximately to 0 ppm in 9 minutes for all the initial propane concentrations. Whereas, the concentration of propane at point 2 takes about 12 minutes to reach to the zero level for all the initial concentrations of propane as can be seen in Figure 7.5 b. In general, the concentration of propane is higher at point 2 than at point 1 because point 1 is located nearer to the air distributor as compared to that of point 2.

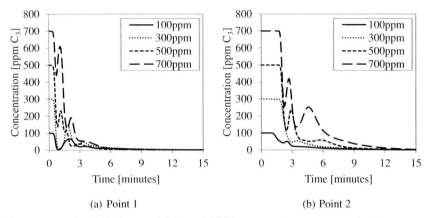

(a) Point 1   (b) Point 2

Figure 7.5:   Case 2 "Influence of different initial in-room pollutant concentrations" – Concentration distribution of propane inside the room

7.3   Case 3 – Influence of different pollutant inlet flowrates

In this study, the initial uniformly distributed concentration of propane present inside the room is zero whereas the pollutant is entering into the room for a certain period of time along with the air flow through the distributor. The influence of different flowrates of the pollutant entering into the room with the constant flow rate of air on the concentration distribution of the pollutant in the room at the ambient temperature and pressure conditions is studied. The purpose of this study is to predict the relative buildup of the pollutant concentration level with the passage of time at the different locations inside the room due to the accidental release of the pollutant into the room at the different flowrates with a constant air flow and then the relative decrease of the pollutant concentration level inside the room with the passage of time due to air flow when the pollutant is stopped entering into the room. The pollutant is entered into the room along with the air for the first 15 minutes. After 15 minutes, only air flow takes place. The total duration of the simulations is 30 minutes with a timestep of 1 s. The three different flowrates of the pollutant entering into the room for the first 15 minutes are 0.0002 kg s^{-1}, 0.0003 kg s^{-1} and 0.0004 kg s^{-1}, whereas after 15 minutes, the flow rate of pollutant is 0 kg s^{-1}. The air flow rate through the distributor is kept constant at 0.34 kg s^{-1}. All the other simulation boundary conditions, parameters and settings are the same as already explained in Chapter 3 and implemented in Chapter 6. The simulation mechanism based on the double configuration explained in Chapter 3 and implemented in Chapter 4 and Chapter 6 is used.

The concentration distributions of propane at different locations inside the room demonstrated in Figure 7.6 generally shows that with the passage of time the propane concentration in the room increases from zero concentration to the maximum value, then stays approximately constant at the maximum value and then decreases sharply as soon as the propane is stopped entering into the room through the distributor. The concentration of propane becomes approximately equal to 0 in 20 minutes and 30 minutes at point 1 and point 2 respectively for all the propane inlet mass flow rates at the constant air flow rate. The rate of decrease of propane concentration is higher at point 1 as compared to that of point 2 because point 2 is located relatively far from the distributor and the fresh air removes the propane faster at point 1 than at point 2. During the propane inlet into room through the distributor, the concentration of propane at a location relatively away from the distributor becomes more uniform than at a location relatively near to the distributor, this is because the air is also entering into the room along with the propane inlet and more turbulence is expected at the locations relatively nearer to the distributor.

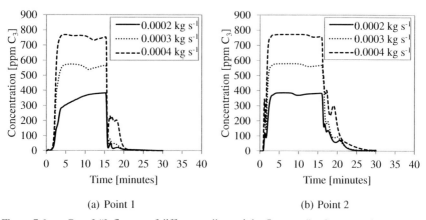

(a) Point 1   (b) Point 2

Figure 7.6:   Case 3 "Influence of different pollutant inlet flowrates" – Concentration distribution of propane inside the room

7.4 Case 4 – Influence of different air inlet flowrates with the same pollutant inlet flowrate

The influence of different flowrates of air with the same flowrate of pollutant entering into the room on the concentration distribution of the pollutant in the room at the ambient temperature and pressure conditions is numerically investigated. In this studied case, the pollutant is entering into the room for a certain period of time along with the air flow through the distributor and the initial uniformly distributed concentration of propane present inside the room is zero. The purpose of this study is to predict the relative buildup of the pollutant concentration level with the passage of time at the different locations inside the room due to the accidental release of the pollutant into the room with the different air flow rates and then the relative decrease of the pollutant concentration level inside the room with the passage of time due to the different air flows when the pollutant is stopped entering into the room. The pollutant is entered into the room along with the air for the first 15 minutes. After 15 minutes, only air flow takes place and the propane inlet is stopped. The total duration of the simulations is 30 minutes with a timestep of 1 s. The simulation mechanism based on the double configuration explained in Chapter 3 and implemented in Chapter 4 and Chapter 6 is used. The four different mass flowrates of air are 0.26 kg s^{-1}, 0.30 kg s^{-1}, 0.34 kg s^{-1} and 0.40 kg s^{-1}, whereas the flowrate of pollutant entering into the room for the first 15 minutes is 0.0003 kg s^{-1} and after 15 minutes is 0 kg s^{-1}. All the other simulation boundary conditions, parameters and settings are the same as already explained in Chapter 3 and implemented in Chapter 6.

The concentration distributions of propane at different locations inside the room are presented in Figure 7.7. The diagram demonstrates that with the passage of time the propane concentration in the room increases from zero concentration to the maximum value, then stays approximately constant at the maximum value and then decreases sharply as soon as the propane is stopped entering into the room through the distributor. The concentration of propane becomes approximately equal to 0 in 30 minutes for all the flow rates of air at point 1. Whereas atpoint 2, the propane concentration is not even reached to zero level in 30 minutes for 0.26 kg s^{-1} of air flow rate. The rate of decrease of propane concentration is higher at point 1 as compared to that of point 2 because point 2 is located relatively far from the distributor and the fresh air removes the propane faster at point 1 than at point 2. The concentration of propane at a location relatively away from the distributor is less fluctuating than at a location relatively near to the distributor, this is because more turbulence is expected at the locations relatively nearer to the distributor. During the propane inlet into the room, the propane reaches to the relatively maximum concentration when the flow rate of air is the least i.e. 0.26 kg s^{-1}. When the propane is stopped entering into the room, the rate of decrease of propane concentration is the largest for the highest mass flow rate of air i.e. 0.40 kg s^{-1}.

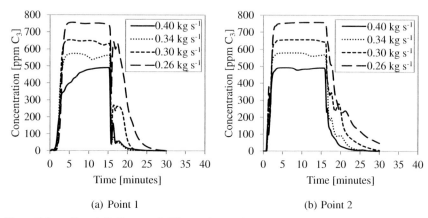

(a) Point 1   (b) Point 2

Figure 7.7:   Case 4 "Influence of different air inlet flowrates with the same pollutant inlet flowrate" – Concentration distribution of propane inside the room

In another study, the application of the developed simulation model to the loss prevention and risk mitigation in the indoor environment is demonstrated when the pollutant is accidentally entered into the room with the air flow. As already mentioned in Section 7.1, the NIOSH Immediately Dangerous to Life or Health (IDLH) concentration of propane is 2100 ppm [87] and the recommended air change rate in the factory buildings, fumes and moisture is 10 – 15 hr^{-1} whereas in the kitchens is 15 – 60 hr^{-1} [86]. The air change rates (ACH) with the corresponding mass flow rates selected for this study are already mentioned in Table 7.2 except for 5 ACH for which the corresponding mass flow rate is 0.0775 kg s^{-1}. It is assumed that the pollutant is accidentally entered into the room along with the air for the first 30 minutes. After 30 minutes, only air flow takes place and the propane inlet is stopped. The flow rate of propane from the cylinder was determined experimentally (demonstrated in Appendix A.2) and is given by the expression $[-(0.0000002[kg\ s^{-2}]*t)+0.0006[kg\ s^{-1}]]$. The total duration of the simulations is 3600 s (60 mins) with a timestep of 1 s. All the other simulation boundary conditions, parameters and settings are the same as already explained in Chapter 3 and implemented in Chapter 6. The simulation mechanism based on the double configuration explained in Chapter 3 and implemented in Chapter 4 and Chapter 6 is used.

The concentration distributions of propane with time at the two different locations inside the room with the different air change rates are demonstrated in Figure 7.8. The two diagrams show that with the passage of time the propane concentrations in the room increase from zero concentration to the maximum values, then decrease due to the time-dependent decreasing flowrate and then decrease relatively faster as soon as the propane is stopped entering into the room through the distributor. The higher the air change rate, the lesser the concentration level of propane in the room for the same flowrate of propane entering into the room and hence the relatively faster rate of removal of propane from the room once the propane is stopped entering into the room. It can be observed from Figure 7.8 that the optimum air change rate, for which the IDLH concentration value of propane (for the period when the propane is being accidentally entered into the room) is approximately below 2100 ppm level, is between 10 to 20 ACH which is exactly in accordance with the recommended air change rate in the factory buildings, fumes and moisture [86].

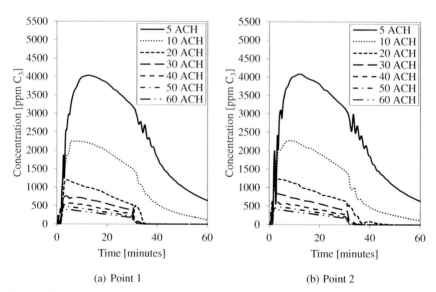

(a) Point 1   (b) Point 2

Figure 7.8:   Case 4 "Influence of different air inlet flowrates with the same pollutant inlet flowrate" – Concentration distribution of propane inside the room

7.5   Case 5 – Influence of different time durations of pollutant inlet

In this study, the pollutant is entering into the room for the certain periods of time along with the air flow through the distributor and the initial uniformly distributed concentration of propane present inside the room is zero. The influence of the different durations of the pollutant inlet into the room with the air flow on the time dependent concentration distribution of the pollutant in the room at the ambient temperature and pressure conditions is studied. The purpose of this study is to predict the relative buildup of the pollutant concentration level at the different locations inside the room due to the accidental release of the pollutant into the room at the different durations of time with the air flow and then the relative decrease of the pollutant concentration level inside the room with the passage of time when the pollutant is stopped entering into the room. The mass flow rates of air and pollutant entering into the room through the distributor are 0.34 kg s^{-1} and 0.0003 kg s^{-1} respectively. The different durations of the pollutant inlet into the room along with the air are 900 s, 1200 s and 1800 s. After these time durations, the flow rate of the pollutant entering into the room becomes zero and only the air flow takes place. The total duration of the simulations is 50 minutes with a timestep of 1 s. All the other simulation boundary conditions, parameters and settings are the same as already explained in Chapter 3 and implemented in Chapter 6. The simulation mechanism based on the double configuration explained in Chapter 3 and implemented in Chapter 4 and Chapter 6 is used.

The concentration profiles of propane at the different locations inside the room for the different time durations of propane inlet into the room exhibited in Figure 7.9 shows that with the passage of time the propane concentration in the room increases from zero concentration to the maximum value, then stays approximately constant at the maximum value and then decreases sharply as soon as the propane is stopped entering into the room through the distributor. When the propane is stopped entering into the room, the concentration of propane at point 2 is generally higher than that at point 1. It can be observed that the rate of decrease of concentration of propane is the same irrespective of the duration of the propane inlet into the room. The concentration of propane in the room becomes steady and constant immediately during the propane inlet into the room. The steady concentration of propane in the room is approximately 575 ppm. The rate of decrease of propane concentration is higher at point 1 than at point 2.

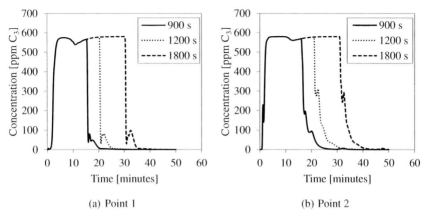

(a) Point 1   (b) Point 2

Figure 7.9:   Case 5 "Influence of different time durations of pollutant inlet" – Concentration distribution of propane inside the room

7.6   Concluding remarks

In this chapter, the application of the developed simulation model to the loss prevention and risk mitigation in the indoor environment was presented in the form of different cases which would otherwise be very difficult to be studied with the help of the expensive and time consuming experiments. In the five studied cases, the study of different phenomena of pollutant spread and distribution inside the room in the presence of air flow and the influence of different parameters on the spatial and temporal in-room distribution of the pollutant entering into the room with the air flow were also demonstrated. It was demonstrated that the developed simulation model could be used as an effective and adequate tool for the loss prevention and risk mitigation in the indoor environment and that there was a significant influence of the different parameters (studied in the five different cases) on the distribution of pollutant with the air flow inside the room.

Chapter 8   Summary and outlook

This chapter summarizes the present work and shows room for improvement of the developed experimental methodology and simulation model for the determination of the distribution of gaseous emissions with the air flow in the room. It also provides an outlook on the further possibilities and prospects of using the developed measurement methodology and simulation model.

8.1   Summary

There are many sources of emissions produced by burning fuel for power or heat, through chemical reactions, and from leaks from industrial processes or equipment. There is always a possibility of a potential hazard when these gases enter into the indoor environment with the air flow. In order to prevent from gas suffocation, explosion and fire accidents, there is a need to observe and understand the different phenomena and the time-dependent spread, propagation and distribution of the leaked gas (e.g. propane) in the presence of air flow inside a room. The determination of the concentration profiles are necessary to evaluate the potential hazard posed by the gas spread. There is also a need to consider the option of performing numerical simulations because experiments can be quite expensive and time-consuming for further detailed studies. Therefore, this work has introduced a continuous measurement system as well as a simulation model based on the application of CFD to determine the spatial and temporal distribution of gaseous emissions with the air flow in the indoor environment.

The main objectives of this work were to develop an appropriate measurement methodology and a 3D CFD transient multicomponent numerical simulation model for the determination of spatial and temporal distribution of gaseous emissions with the air flow in the indoor environment. This work was also aimed at comparing the numerical simulation results of different CFD programs for a 2D base case model of indoor air flow with and without emission source under laminar and turbulent flow conditions for the purpose of developing a better basic physical understanding and for the selection of the suitable and appropriate CFD program for the further development of the simulation model enabling the determination of the temporal and spatial distribution of gaseous pollutants with the air flow inside the room. The ultimate goal was to develop and validate the numerical simulation model for the determination of gaseous emissions in the indoor environment which could be used as an effective and adequate tool for the loss prevention and risk mitigation in the industrial indoor environment and for the design of a safer indoor environment. One of the goals was also to apply the developed simulation model to the loss prevention and risk mitigation in the indoor environment and to study the influence of different parameters on the distribution of gaseous emissions in the presence of air flow in the indoor environment so as to minimize the expensive and time consuming experimentation efforts.

Experimental

An experimental setup was developed, the measurement technique was selected and several experimental runs were conducted first involving the two components (propane and air) and then the three components (propane, CO_2 and air) for the purpose of investigation of the spread and distribution of the gaseous emissions in the room with the air flow and for the experimental validation of the developed 3D numerical simulation model based on double configuration. The flame ionization detectors and infrared photometers were employed to determine the time-dependent concentrations of propane and CO_2 respectively at different locations, whereas the flowrate of the gases was measured with the help of Wilson flow grid. The experiments were carried out with and without flow obstruction present inside the room. All the experimental runs were carried out at ambient temperature and pressure conditions. For the case without flow obstruction present inside the room, it has been revealed that the concentration of the component at each relative measurement position placed inside the room was generally almost the same relative to each other because there was no physical hindrance present inside the room and the gas flow was assumed to be uniformly distributed into the room through the distributor. The concentration of the component at all the three measurement points was first increased, reached the maximum level, then slightly decreased with time and ultimately fallen to a minimum level immediately after the gaseous components were stopped entering into the room.

The experimental investigation of the influence of the flow obstruction present in the room on the distribution of the gaseous emission showed that the obstruction present in the way of the flow had a significant influence on the concentration profiles of the gaseous component inside the room. The investigation of the impact of the flow from the distributor on the concentration distribution of the propane at the top and bottom measurement points located at the same distance from the distributor when the propane was entering into the room demonstrated that the concentration of propane at the point placed at a relatively higher position was greater than that at the point placed at the relatively lower position. When the propane inlet was stopped, the propane concentration at the point placed at a relatively higher position was lower than that at the point placed at the relatively lower position. The comparison of the concentrations of propane at the locations nearer and away from the distributor with the passage of time when the propane was entered into the room with air showed that the concentration of the propane at a point placed nearer to the distributor (from where the propane along with air is entering into the room) was generally higher than at the point placed relatively away from the distributor. It was also revealed that the pollutant concentration at a point placed relatively nearer to the distributor was more fluctuating than that at the point placed away from the distributor because of the more turbulent air flow nearer to the distributor.

Simulation

For the purpose of developing a better physical understanding and for the selection of the suitable and appropriate CFD program for the development of the 3D transient multicomponent simulation model based on the mechanism double configuration which enables the determination of the temporal and spatial distribution of gaseous emissions with the air flow inside the room, the comparison of the numerical simulation results of different CFD programs for a 2D base case model of indoor air flow with and without emission source under laminar and turbulent flow conditions was done. It has been seen that all the three employed CFD codes had almost the equal capabilities to model the indoor air flow, and the simulation results of all the three programs were in good agreement against each other for both the laminar flow and the turbulent flow. The comparison of the numerical simulation results and the relative capabilities of CFX and FLUENT to model and determine the spatial distribution of formaldehyde emission in the presence of air flow under both the laminar flow condition and the turbulent flow condition in the room shows that the results from both the CFD programs were in good agreement with each other under both laminar flow and turbulent flow conditions. CFX was selected for the development of the 3D transient multicomponent numerical simulation model based on the double configuration and for the further advanced simulation work because of its availability for this research work and the ease of implementation of the double configuration mechanism using configuration control and the complex boundary conditions with the user defined expressions and functions.

The 3D transient multicomponent numerical simulation model to determine the spatial and temporal distribution of gaseous emissions inside the room with air flow was developed based on the mechanism of double configuration using ANSYS CFX 15.0 in which the steady state and transient analyses had been sequenced in such a way that the transient simulation automatically starts upon the completion of the steady state simulation. This was achieved by setting up the configurations of these analyses in a subsequent step by using the configuration control in ANSYS CFX 15.0. The first configuration was a steady-state analysis involving only air flow and no component/s inlet into the room through the distributor. The second configuration was a transient-state analysis involving the component/s inlet into the room along with the air through the distributor. The second configuration took the setup of the steady state and adapted it for a transient analysis, hence the results from the steady-state analysis were used as the initial guess for the transient analysis.

The validation of the developed 3D multicomponent numerical simulation model based on the mechanism of double configuration for the two components system and the three components system was done against the experimental data. The boundary conditions and model parameters were transferred from the experiment to the simulation. The simulation results were in good agreement against the experimental data. The difference between simulation and experimental results was not more than 20 percent which is an acceptable standard for the CFD validations. The simulation and experimental results generally showed the same response with time, however the simulated propane concentration at all the measurement points immediately reached to the zero concentration level relatively fast after the propane inlet into the room was stopped because, contrary to the reality, the simulation was carried out at the ideal conditions in which the pollutants were efficiently removed out of the room by the efficient air flow once the pollutant was stopped entering into the room. The simulated propane concentration was not fluctuating because the propane was efficiently distributed uniformly into the room by the simulated ideal air flow. The simulation model was validated against the experimental data for both the cases with and without flow obstruction placed inside the room. The evaluation of different turbulence models to model the distribution of gaseous emissions inside the room with the air flow demonstrated that all the employed turbulence models predict the same concentration distribution compared against the experimental data. The k-ε model was employed in the present work.

The further simulation studies were carried out for the purpose of application of the developed simulation model to the loss prevention and risk mitigation in the indoor environment and to study the influence of different parameters on the pollutant spread and distribution inside the room in the presence of air flow which would otherwise be very difficult to be studied with the help of the expensive and time consuming experiments. It has been demonstrated that the developed simulation model could be used as an effective and adequate tool for the loss prevention and risk mitigation in the indoor environment. It has also exhibited that there was a significant influence of the different parameters (studied in the five different cases) on the distribution of pollutant in the presence of air flow inside the room.

8.2   Outlook

Due to the limited availability of FIDs for this research work, only three FIDs were employed at one time to measure the concentrations of propane at the different locations of the room. For even better and complete understanding of the distribution of pollutants inside the room with the air flow, the concentrations of propane should be measured at larger number of measurement locations and therefore, a large number of FIDs are suggested to be employed for this purpose. The same is true for the carbon dioxide measurement. A large number of infrared photometers are recommended to be employed to measure the concentrations of carbon dioxide at the different locations inside the room to have a complete physical understanding. Moreover, the measurements and simulations in the present work were performed with and without flow obstruction inside the room. Only one flow obstruction was placed inside the room for the measurements and simulations. To determine the distribution of pollutants inside the room with the air flow under the more realistic situations, it is recommended to place more number of flow obstructions inside the room.

It is recommended to extend the developed simulation model to accommodate the non-isothermal conditions. The comparison of different CFD programs for the indoor air flow with and without emission source placed inside the room under non-isothermal conditions should also be done for the purpose of selection of the suitable and appropriate CFD program for the indoor non-isothermal systems.

The developed simulation model based on double configuration in CFX is suggested to be regenerated in OpenFOAM and FLUENT for the purpose of comparison of the simulation results of different CFD programs for the problem of gaseous emissions in indoor environment. The use of open source CFD package OpenFOAM should be encouraged. The implementation of double configuration in OpenFOAM is considered to be a big challenge. In double configuration, the steady state and transient analyses are sequenced in such a way that the transient simulation automatically starts upon the completion of the steady state simulation and the results from the steady-state analysis are used as the initial guess for the transient analysis. In OpenFOAM, there is no possibility to set up the configurations of these analyses in a subsequent step with the help of the configuration control. The advanced programming skills would be required to implement this approach of double configuration in OpenFOAM.

The development of an experimental methodology for the simultaneous measurement of gaseous and particulate emissions at different locations inside the room with the air flow is considered as an important future work. Parallel to this, the development and experimental validation of the simulation model for the simultaneous prediction of gaseous and particulate emissions inside the room should also be done.

The developed simulation model based on the double configuration has a potential to be employed for the design of the clean rooms and further scenarios of the loss prevention and risk mitigation in the industries. It is also suggested to employ the modeling concept developed in the present work to simulate the distribution of VOCs emissions in cars. Moreover, the development of a measurement methodology to measure the temporal and spatial concentration distribution of VOCs inside the vehicle cabin under different conditions is also seen as a potential future research work.

The developed simulation model in this work uses RANS turbulence modeling approach to model the distribution of gaseous pollutants inside the room with the air flow. It is also suggested to regenerate the same developed model using Large Eddy Simulation (LES) instead of RANS turbulence modeling for the purpose of capturing the small length scales.

Appendix A   Experimental

A.1 Measurement technique

This section explains the technical functioning of the measurement technique used in this work. First, the operation and technical detail of the flame ionization detector (FID) employed to determine the propane concentrations at different locations inside the room are described. Then the working operation of the infrared photometer used for the determination of the carbon dioxide concentrations at different measurement positions inside the room is presented. At the end of this section, the measurement principle and the technical detail of the Wilson flow grid used for the detection of volume flow rate of air are explained.

A.1.1 Flame ionization detector

In the present work, the flame ionization detectors (FIDs) from the company SK-Elektronik GmbH [77] have been employed to determine the time dependent concentrations of propane at different locations inside the room. The working operation of FID is schematically shown in Figure A.1.

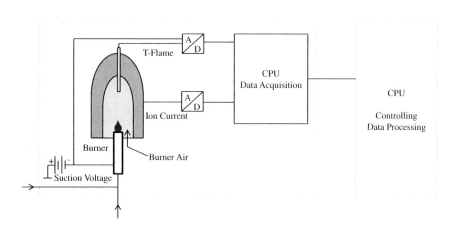

Figure A.1: Schematic description of the operation of the employed Flame Ionization Detector (FID) [77]

The gas sample is continuously taken into FID and the negative ions are created when the organic components are burnt in a hydrogen flame. A flame is created by burning the pure hydrogen and cleaned air (burner air) in a temperature controlled burning chamber. A negative suction voltage is maintained across the burner nozzle and an electrode to measure the produced negative ions. The existing current is detected by a highly sensitive current amplifier. The amplified signals are digitized directly at the detector and get transmitted as digital data to the CPU for the analysis. The current measured is directly proportional to the content of org. C in the sample/burning flame. The technical data of the flame ionization detector used in this work is presented in Table A.1.

Table A.1: Technical data of the employed Flame Ionization Detector from the company SK-Elektronik GmbH [78]

Measurement range	0.5 mg org.C m^{-3} to 100000 mg org.C m^{-3}
Detection limit	< 0.01 mg org.C m^{-3}
Linearity	$< 4\%$ above measurement range
Response time (T90)	< 0.5 s if measured value > 20 mg org.C m^{-3} < 5.0 s if measured value < 20 mg org.C m^{-3}
Samplegas (self-drawed)	ca. 25 L h^{-1} with 1013 hPa alternative ca. 2 L h^{-1}; ca. 5 L h^{-1}; ca. 60 L h^{-1}
Samplegas-pressure	800 mbar to 1600 mbar
Fuel gas	Hydrogen 0.7 bar to 1.0 bar / < 80 mL min^{-1} Quality 5.0
Burner air	Over internal catalyst or optional Synthetic air 1.0 bar to 1.5 bar
Measured value output	0(4) mA to 20 mA
Ambient temperature	-5 °C to 40 °C
Detector-temperature	T2: 110 °C to 170 °C T3: 130 °C to 165 °C T4: 95 °C to 100 °C
Catalyst-temperature	400 °C

A.1.2 Infrared photometer

In the present work, the infrared photometers Uras26 from the company ABB Automation GmbH [79] have been employed to determine the time dependent concentration of carbon dioxide at different locations inside the room. Figure A.2 shows the schematic working operation of a typical infrared photometer [88]. The measurement principle of the Uras26 is based on the capability of gas molecules to specifically absorb infrared (IR) radiation [89]. Depending on the gas concentration, the energy of a light beam is removed within a certain frequency range. The sample of interest is held in the gas-filled opto-pneumatic detectors. The gas absorbs the radiant energy which causes a change in temperature and thereby a change in pressure in the detector. An electrical signal is produced via a membrane capacitor due to this change in pressure. An extremely high selective sensitivity with regard to gases such as CO, CO_2, SO_2, NO, CH_4 and N_2O (in our case CO_2) is provided by the correlation between the detector gas and the sample gas. The periodic modulation of the IR radiation source with a chopper wheel and subsequent frequency and phase selective amplification provide a low-noise measurement. The technical data and the measurement range limits of the Infrared Photometer Uras26 used in this work are given in Table A.2 and Table A.3 respectively. The analyzer has one physical measurement range per sample component, and the measurement ranges are freely adjustable within the limits as shown in Table A.3.

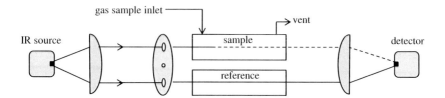

Figure A.2:   Schematic description of the operation of a typical Infrared Photometer [88]

Table A.2:   Technical data of the Infrared Photometer Uras26 from the company ABB
              Automation GmbH [80]

Wavelength range	2.5 – 8 µm
Measurement Range Limits	See Table A.3
Linearity Deviation	≤ 1 % of span
Detection Limit	≤ 0.4 % of span
Inlet Pressure	2 – 500 hPa
Outlet Pressure	Atmospheric pressure
Flow Rate	$20 - 100 \text{ L h}^{-1}$

Table A.3:   Measurement range limits of Infrared Photometer Uras26 from the company
              ABB Automation GmbH [80]

Smallest Measurement Range	Largest Measurement Range
0 – 100 ppm	0 – 500 ppm
0 – 200 ppm	0 – 1000 ppm
0 – 600 ppm	0 – 3000 ppm
0 – 2000 ppm	0 – 10000 ppm
0 – 0.6 Vol.-%	0 – 3 Vol.-%
0 – 2 Vol.-%	0 – 10 Vol.-%
0 – 6 Vol.-%	0 – 30 Vol.-%
0 – 20 Vol.-%	0 – 100 Vol.-%

A.1.3   Wilson flow grid

In the present work, the Wilson flow grid from the company Halstrup-Walcher [81] has been
employed to determine the differential pressure (hence, the volumetric flow rate) of the gas in
the duct. The flow grid consists of 6 tubes arranged radially around a central hub. The sum of
the static and dynamic pressure is measured and recorded by the Wilson flow grid at several
points and the static pressure is subtracted [90]. The values of the conveyed volume flow and
the air exchange rate are then calculated. The technical data of the Wilson flow grid employed
in this work is mentioned in Table A.4.

Table A.4:   Technical data of the employed Wilson flow grid [82]

Measurement range	0 – 100 Pa
Accuracy	± 0.2 % or ± 0.5 % of the scaled range (40 – 100 % of max. value) (min. 0.3 Pa)
Medium	Air, non-aggressive gases
Operating temperature	+10 °C to +50 °C
Sensor response time	25 ms
Rated input	approx. 6 W

A.2 Weight analysis of propane cylinder

The weight of the propane cylinder was measured with time with the help of a weight balance and a stop watch for 30 minutes with an interval of 30 seconds when the propane was flowing out from the cylinder. The pressure of the propane cylinder and the initial weight of propane in the cylinder (when new) were 7.4 bar and 11 kg (22 kg = 11 kg propane gas + 11 kg empty cylinder) respectively. The valve of the propane cylinder was opened in such a way that the maximum flow of propane takes place. Figure A.3 shows the weight vs. time relationship of the propane cylinder when the propane was flowing out from the cylinder.

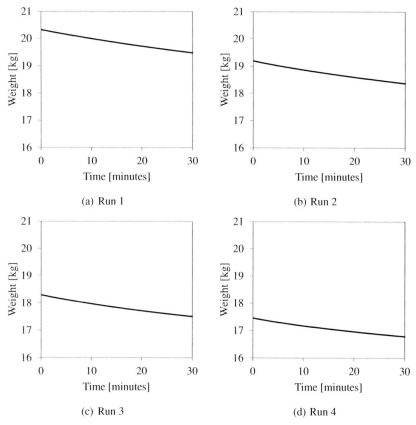

(a) Run 1

(b) Run 2

(c) Run 3

(d) Run 4

Figure A.3:   Weight vs. time relationship of the propane cylinder for all the four experimental runs when the propane was flowing out from the cylinder

In total, four experimental runs were performed. It is important to note that the weight of the propane cylinder presented here will be the sum of the weight of the cylinder (when empty) and the gas inside it. For the first experimental run, the weight of the propane cylinder before the start of the experiment was 20.322 kg. The total propane gas used for run 1, run 2, run 3 and run 4 for 30 minutes duration each was 0.846 kg, 0.822 kg, 0.794 kg and 0.670 kg respectively. Hence, the quantity/amount of the propane released in 30 minutes for second experimental run was less than the first experimental run. The same behavior was observed for third and fourth experimental runs. It was noted that the weight vs. time relationship was linear. As exhibited in Figure A.4, it was found out that the flow rate of the propane releasing from the cylinder was decreasing with time, and this relationship was linear.

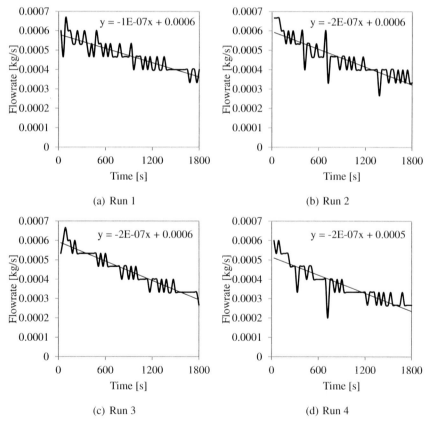

(a) Run 1

(b) Run 2

(c) Run 3

(d) Run 4

Figure A.4:   Flowrate vs. time relationship of propane flowing out from the cylinder for all the four experimental runs

A.3   Further experiments

In this section, the spread of an accidentally released gaseous pollutant in the presence of air flow in the room is described experimentally in the form of different case scenarios to investigate the different phenomena related to the pollutant spread in the room. Three FIDs were employed for the measurement of the concentrations of propane at different locations of the room.

A.3.1   Two components system (without flow obstruction)

Case – Measurement points away from the distributor (z = 1.125 m)

In this case of two components system without flow obstruction present in the room, the concentration of the accidentally released propane with air into the room is examined with time at the measurement points located away from the distributor. When the steady flow conditions are maintained, the propane is fed into the room along with air for 30 minutes. The experiment is carried out at room temperature and pressure conditions. After 30 minutes, the propane inlet into the room is stopped and only the air flow takes place. The total duration of the experiment is 48 minutes. The volume flow rate of the gases (air and pollutant) from the distributor is adjusted at 280 – 300 L s^{-1}. The concentration distribution of propane at all the measurement points is demonstrated in Figure A.5.

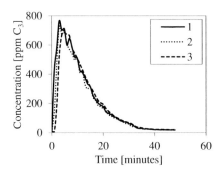

Figure A.5:   Two components system (without flow obstruction) – Concentration distribution of propane inside the room at the three measurement points located away from the distributor (z = 1.125 m)

The coordinates and locations of the measurement points for this case have already been tabulated and shown in Table 5.2 and Figure 5.8 respectively. As exhibited in Figure A.5, the propane concentration at all the three measurement points is first increased, reached the maximum level, then slightly decreases with time and ultimately falls to a minimum level immediately after the propane is stopped entering into the room. The concentration of propane reaches to a level of approximately 700 ppm in the first 5 minutes at almost all the measurement locations. The concentration of propane at all the three measurement positions is generally approximately the same because there is no physical hindrance present inside the room (no fireplace present inside the room). There are fluctuations in the values of the measurements because of the turbulent air flow and the non-uniform inlet flow of the propane. At 30 minutes, the concentration of propane is 61 ppm, 58 ppm and 67 ppm at points 1, 2 and 3 respectively. The propane concentration decreases with time even when the propane is entering into the room because the flowrate of propane entering into the room from the cylinder is decreasing with time. At the end time of 48 minutes, the concentration at all the measurement points reduced close to 18 ppm.

A.3.2 Two components system (with flow obstruction)

Case 1 - Relative distance

In this case, the concentration of pollutant in the room in the presence of air flow at two different points located at the same height but at different distances from the distributor is measured before and after the pollutant is stopped entering into the room at the ambient temperature and pressure conditions. The objective is to compare the concentrations at locations nearer and away from the distributor with the passage of time when the pollutant is accidentally released into the room with air. The Table A.5 and Figure A.6 show the coordinates and locations respectively of the measurement points for this case.

Table A.5:   Two components system (with flow obstruction) – Coordinates of the measurement points for case 1 "Relative distance"

Measurement point 1	Measurement point 2	Measurement point 3
outlet	(2.050, 1.200, 1.125)	(2.050, 1.200, 3.375)

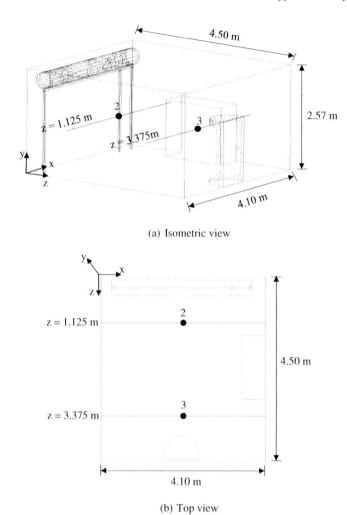

(a) Isometric view

(b) Top view

Figure A.6:   Two components system (with flow obstruction) – Locations of the
measurement points for case 1 "Relative distance"

The measurement location "outlet" is situated at the exhaust duct system on the roof of the chamber as shown in Figure 5.3 a of Chapter 5. In Figure A.7 a, the propane concentration profiles at the three measurement locations are shown. The propane is injected into the room along with air for 30 minutes when the steady flow conditions are maintained. The propane inlet into the room is stopped after 30 minutes, and only the air flow takes place. The volume flow rate of the air and pollutant from the distributor is maintained at $270 - 290$ L s^{-1} by adjusting the speed of the tube fan. The total duration of this experiment is 105 minutes. The propane concentration at all the three measurement points is first increased, reached the maximum level, then slightly decreases with time and ultimately falls to a minimum level immediately after the propane is stopped entering into the room. The concentration of propane reaches to a level of 1143 ppm, 999 ppm and 1190 ppm at points 1, 2 and 3 in 7 minutes respectively. The concentration of propane at point 3 is generally higher than that at point 2. The point 3 is placed relatively nearer to the distributor from where the pollutant along with air is entering into the room and therefore the concentration at point 3 is more fluctuating than that at point 2 because of the more turbulent air flow nearer to the distributor. At 40 minutes, the concentration of propane is about 73 ppm, 76 ppm and 43 ppm at points 1, 2 and 3 respectively. As shown in Figure A.7 b, at 60 minutes the concentration of propane is 29.42 ppm, 32.51 ppm and 28.76 ppm at points 1, 2 and 3 respectively. When the propane inlet is stopped, the propane concentration at point 3 (nearer to the distributor) is slightly lower than that at point 2 (away from the distributor). This is because the air flow is higher and more turbulent at the point nearer to the distributor than the point away from the distributor hence resulting in the relatively faster rate of decrease in the concentration of propane at the point nearer to the distributor than at the point away from the distributor. At the end time of 105 minutes, the concentration at all the measurement points reduced close to 0 ppm.

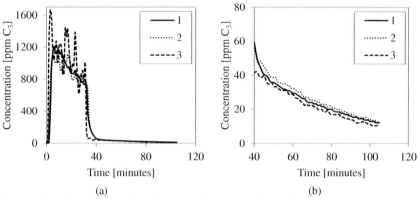

Figure A.7: Two components system (with flow obstruction) – Concentration distribution of propane inside the room for case 1 "Relative distance"; (a) Total period, (b) For a period from 40 minutes to 106 minutes

Case 2 - Relative height

In this last case, the concentration of pollutant (propane) in the room at the top and bottom measurement points located at the same distance from the distributor is measured with respect to time when the pollutant is accidentally released into the room with air. The objective is to investigate the impact of the flow from the distributor on the concentration distribution of the pollutant at the top and bottom measurement points located at the same distance from the distributor before and after the pollutant is stopped entering into the room. The total duration of the experiment is 113 minutes and the experiment is carried out at room temperature and pressure conditions. The Table A.6 and Figure A.8 show the coordinates and locations respectively of the measurement points for this case. The measurement location "outlet" is situated at the exhaust duct system on the roof of the chamber as shown in Figure 5.3 a of Chapter 5.

Table A.6:   Two components system (with flow obstruction) – Coordinates of the measurement points for case 2a "Relative height"

	Measurement point 1	Measurement point 2	Measurement point 3
Case 2a	outlet	(1.025, 2.000, 1.125)	(1.025, 0.100, 1.125)

When the steady flow conditions are maintained, the propane is fed into the room along with air for 30 minutes. After 30 minutes, the propane inlet into the room is stopped and only the air flow takes place. The volume flow rate of the gases from the distributor is maintained at $270 - 280$ L s^{-1}. In Figure A.9 a, the propane concentration profiles at the 3 FID sensors are shown. The concentration of propane reaches to a level of 1068 ppm, 1204 ppm and 808 ppm at points 1, 2 and 3 in 10 minutes respectively. It ultimately falls to a minimum level immediately after the propane is stopped entering into the room. The concentration of propane at point 2 is higher than that at point 3. The point 2 is placed at a higher position near to the exhaust than that of point 3 which results in better air flow at the top (point 2) than at the bottom (point 3).

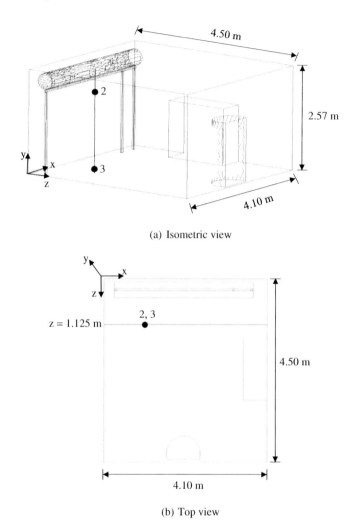

(a) Isometric view

(b) Top view

Figure A.8:   Two components system (with flow obstruction) – Locations of the measurement points for case 2a "Relative height"

At 80 minutes the concentration of propane is 5.71 ppm, 3.39 ppm and 4.75 ppm at points 1, 2 and 3 respectively as shown in Figure A.9 b. When the propane inlet is stopped, the propane concentration at point 2 is lower than that at point 3. This is because the propane is denser than air and therefore the bottom section contains more propane than the top section. There are fluctuations in the values of the measurements because of the turbulent air flow and the non-uniform inlet flow of the propane. The propane concentration decreases with time even when the propane is entering into the room, this is because the flowrate of propane entering into the room from the cylinder is decreasing with time. At the end time of 113 minutes, the concentration at all the measurement points reduced close to 0 ppm.

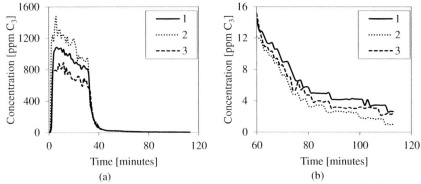

Figure A.9: Two components system (with flow obstruction) – Concentration distribution of propane inside the room for case 2a "Relative height"; (a) Total period, (b) For a period from 60 minutes to 113 minutes

In one more experiment (case 2b), the concentration of propane inside the room is determined at the two measurement points located at the middle of the room at different heights. The Table A.7 and Figure A.10 show the coordinates and locations respectively of the measurement points for this case. The measurement location "outlet" is situated at the exhaust duct system on the roof of the chamber as shown in Figure 5.3 a of Chapter 5.

Table A.7: Two components system (with flow obstruction) – Coordinates of the measurement points for case 2b "Relative height"

	Measurement point 1	Measurement point 2	Measurement point 3
Case 2b	outlet	(2.050, 2.000, 2.250)	(2.050, 0.100, 2.250)

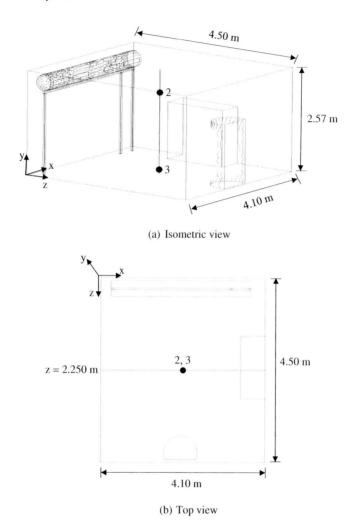

(a) Isometric view

(b) Top view

Figure A.10: Two components system (with flow obstruction) – Locations of the measurement points for case 2b "Relative height"

The propane concentration profiles at the two measurement locations inside the room and one at the outlet are exhibited in Figure A.11. This experiment is also carried out at room temperature and pressure conditions and the total duration of this experiment is 65 minutes. The flow rate of the gases (air and pollutant) from the distributor is adjusted at $270 - 280$ L s^{-1}. The propane is entered into the room along with air for 30 minutes when the steady flow conditions are assumed to be maintained with the initial air flow. After 30 minutes, the propane inlet into the room is stopped and only the air flow takes place. The concentration of propane reaches to a level of 768 ppm, 911 ppm and 600 ppm at points 1, 2 and 3 in 10 minutes respectively. It ultimately falls to a minimum level immediately after the propane is stopped entering into the room. The concentration of propane at point 2 is generally higher than that at point 3. The point 2 is placed at a higher position near to the exhaust than that of point 3 which results in better air flow at the top (point 2) than at the bottom (point 3). At 40 minutes, the concentration of propane is 45 ppm, 37 ppm and 46 ppm at points 1, 2 and 3 respectively. When the propane inlet is stopped, the propane concentration at point 2 is lower than that at point 3 because the propane is denser than air and therefore the bottom section contains more propane than the top section. The concentration at all the measurement points reduced close to 25 ppm at the end time of 65 minutes. Because of the turbulent air flow and the non-uniform inlet flow of the propane, some fluctuations in the values of the measurements are observed.

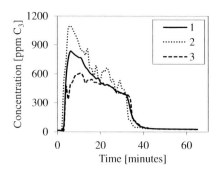

Figure A.11: Two components system (with flow obstruction) – Concentration distribution of propane inside the room for case 2b "Relative height"

Appendix A: Experimental

Appendix B   Simulation

B.1   2D base case model

In this section, the numerical simulation of indoor air flow under laminar flow condition is first carried out using three CFD programs (CFX, FLUENT and OpenFOAM) followed by the simulation of the same model with an emission source placed inside the room using two CFD programs (CFD and FLUENT). At the end, the results of OpenFOAM running in parallel on the computing cluster are discussed.

B.1.1   Laminar indoor air flow model

The comparison of the numerical simulation results of CFX, FLUENT and OpenFOAM to model the indoor air flow under laminar flow condition is presented in this section. The sketch of this model has already been shown in Figure 3.1 of Chapter 3. The simulation settings and boundary conditions have been implemented in accordance with Table B.1. The inlet velocity perpendicular to the inlet surface is $5.56 \cdot 10^{-3}$ m s^{-1}. The whole model is under atmospheric pressure, thus the relative pressure at the outlet is 0 Pa. The laminar indoor air flow is modelled under isothermal condition (T = 338.15 K).

The velocity (absolute) contours from the three CFD programs are demonstrated in Figure B.1. It can be seen that the basic form of the flow obtained from the three CFD programs is substantially the same. CFX and OpenFOAM give the slightly higher velocity values at the middle lower region of the room as compared to that of FLUENT.

The locations of the four lines (x = 3.0 m, x = 6.0 m, y = 0.084 m and y = 2.916 m), where the velocity distributions from all the three CFD programs are compared, have already been displayed in Figure 4.2. The simulation results for the non-dimensional velocities from CFX, FLUENT and OpenFOAM along the four lines (x = 3.0 m, x = 6.0 m, y = 0.084 m and y = 2.916 m) defined in Figure 4.2 are shown in Figure B.2. It should be noted that the velocity in the x-direction is dimensioned by the entry velocity. As can be seen in Figure B.2, the simulation results of the three CFD programs for the indoor air flow under isothermal laminar flow condition are generally in good agreement with each other.

Table B.1:   Indoor air flow (laminar) – Simulation settings and boundary conditions
implemented in CFX, FLUENT and OpenFOAM

Settings and boundary conditions	CFX	FLUENT	OpenFOAM
Simulation type	Steady state	Steady state	Steady state
Fluid information	Air	Air	Air
Solution methods	High Resolution	SIMPLE	SIMPLE
Domain	Reference Pressure: 101325 Pa Isothermal (T = 338.15 K) Laminar	Operating Pressure: 101325 Pa Isothermal (T = 338.15 K) Laminar	Reference Pressure: 101325 Pa Isothermal (T = 338.15 K) Laminar
Boundary conditions			
Velocity-Inlet	normal speed: $5.56 \cdot 10^{-3}$ m s^{-1}	magnitude, normal to BC: $5.56 \cdot 10^{-3}$ m s^{-1}	Fixed value: $5.56 \cdot 10^{-3}$ m s^{-1}
Pressure-Outlet	Average Static Pressure Static Pressure: 0 Pa	Gauge Pressure: 0 Pa	Fixed value: 0 Pa
Wall	No slip wall	No slip wall, Stationary Wall	No slip wall
Symmetry	Symmetry	-	Symmetry

Figure B.2 a and b show the comparison of the simulation results of the three CFD programs
along the two lines at x = 3.0 m and x = 6.0 m respectively. The simulation results are relatively in better agreement at x = 6.0 m as compared to those at x = 3.0 m. A good agreement
is obtained at the line x = 3.0 m between y = 0.5 to 2.3 m. The non-dimensional velocity
from FLUENT deviates larger at the line x = 3.0 m between y = 0 m to 0.5 m. The non-
dimensional velocities remain almost identical at the line x = 6.0 m. The comparison of the
numerical simulation results of the three programs along the two lines at y = 0.084 m and
y = 2.916 m is exhibited in Figure B.2 c and d respectively.

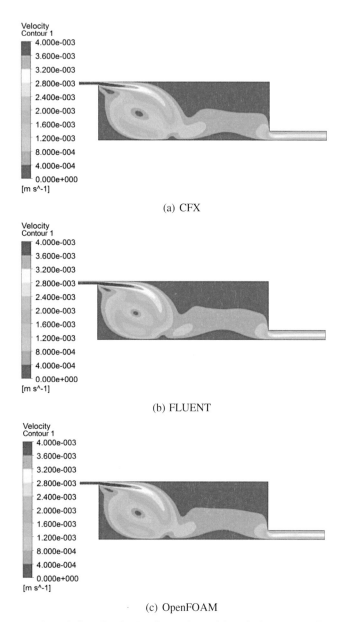

(a) CFX

(b) FLUENT

(c) OpenFOAM

Figure B.1:   Indoor air flow (laminar) – Comparison of the velocity contours from the three CFD programs

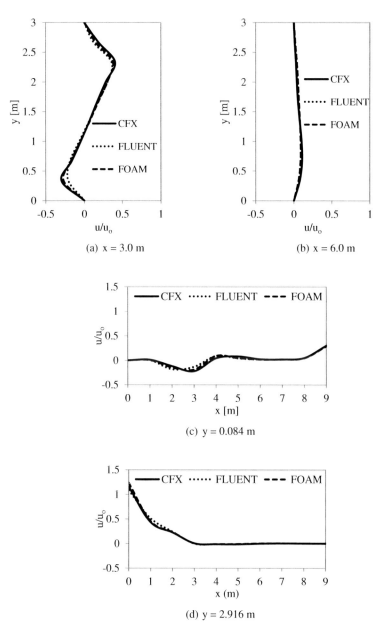

(a) x = 3.0 m

(b) x = 6.0 m

(c) y = 0.084 m

(d) y = 2.916 m

Figure B.2:   Indoor air flow (laminar) – Comparison of numerical simulation results (non-dimensional velocities) of the three CFD programs at the different lines locations (x = 3.0 m, x = 6.0 m, y = 0.084 m and y = 2.916 m)

The deviation among the values of the non-dimensional velocities from the three CFD programs is relatively higher at the line y = 0.084 m between x = 1 m to 6 m as can be seen in Figure B.2 c. At the line y = 0.084 m after x = 6 m, the non-dimensional velocities from the three programs are almost identical. As shown in Figure B.2 d, at the line y = 2.916 m the deviations among the non-dimensional velocities from the three programs are slightly significant from x = 0 m to 2 m, but the velocities become almost identical after x = 2 m. The velocity values at all the walls are zero because of the no-slip condition at the walls.

B.1.2   Laminar indoor air flow model with an emission source

The comparison of the simulation results of CFX and FLUENT for the determination of the spatial distribution of formaldehyde emission in the presence of air flow under laminar flow condition at steady state in the room is presented in this section. The sketch of the model with an emission source has already been demonstrated in Figure 4.7. The simulation settings and boundary conditions implemented in CFX and FLUENT are shown in Table B.2. All the boundary conditions and simulation settings are the same as presented in section B.1.1 for the isothermal laminar indoor air flow except an introduction of an emission source as a new source term inserted in the mass balance. As the changes in the mass balance have an influence on the other conservation laws, so the velocities and the mass fractions from both the CFD programs are compared. The formaldehyde emission rate is $3.475 \cdot 10^{-10}$ kg m^{-3} s^{-1}.

The comparison of the velocity (absolute) contours from the two CFD programs exhibited in Figure B.3 shows that the velocities predicted by both the programs are almost identical. The velocity is the minimum at the top right and bottom left corners of the room. The locations of the four lines (x = 3.0 m, x = 6.0 m, y = 0.084 m and y = 2.916 m), where the velocity distributions from both the CFD programs are compared, have already been demonstrated in Figure 4.2. The dimensionless velocities from the two programs for the laminar indoor air flow with an emission source at the two horizontal (y = 0.084 m and y = 2.916 m) and the two vertical lines (x = 3.0 m, x = 6.0 m) are presented in Figure B.4.

Table B.2: Indoor air flow (laminar) with emission source – Simulation settings and boundary conditions implemented in CFX and FLUENT

Settings and boundary conditions	CFX	FLUENT
Simulation type	Steady state	Steady state
Fluid information	Mixture	Mixture
Solution methods	High Resolution	SIMPLE
Domain	Reference Pressure: 101325 Pa Isothermal (T = 338.15 K) Laminar	Operating Pressure: 101325 Pa Isothermal (T = 338.15 K) Laminar
Boundary conditions		
Velocity-Inlet	normal speed: $5.56 \cdot 10^{-3}$ m s^{-1}	magnitude, normal to BC: $5.56 \cdot 10^{-3}$ m s^{-1}
Pressure-Outlet	Average Static Pressure Static Pressure: 0 Pa	Gauge Pressure: 0 Pa
Wall	No slip wall	No slip wall, Stationary Wall
Symmetry	Symmetry	Symmetry
Emission source	CHOH: $3.475 \cdot 10^{-10}$ kg m^{-3} s^{-1}	CHOH: $3.475 \cdot 10^{-10}$ kg m^{-3} s^{-1}

As shown in Figure B.4 a, the highest velocity at the line x = 3.0 m is obtained at y = 2.7 m which corresponds to nearly the same horizontal position of the inlet. When considering the velocity at the line x = 6.0 m as exhibited in Figure B.4 b, there are two regions with a relatively higher velocity, one at y = 1.9 m and second at y = 0.35 m. At the line y = 0.084 m, the velocity is nearly zero from x = 0 m to 4 m and then relatively increases near the outlet as shown in Figure B.4 c. At the line y = 2.916 m, the velocity is higher near the inlet and decreases subsequently near to zero with the flow of the fluid in the x direction as demonstrated in Figure B.4 d.

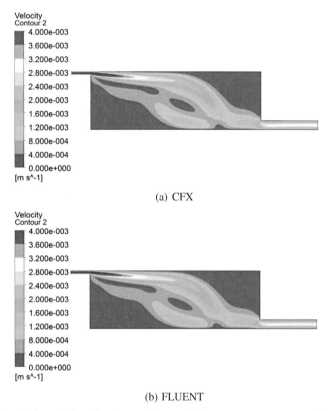

(a) CFX

(b) FLUENT

Figure B.3: Indoor air flow (laminar) with an emission source – Comparison of the velocity contours from the two CFD programs

The concentration distribution of formaldehyde in terms of mass fraction in the room in the presence of air flow under laminar flow condition at the steady state from both the CFD programs is presented in Figure B.5. As depicted in Figure B.5, there is almost no difference in the mass fraction distribution between the two programs. The spread of formaldehyde in the room is influenced by the air flow. On the right side of the source, in the vicinity of the right wall, the propagation is curved due to the influence of the circulation. Due to the low emission rate and the slow flow velocity (laminar), a limited spread of formaldehyde is occurred in the whole calculation area.

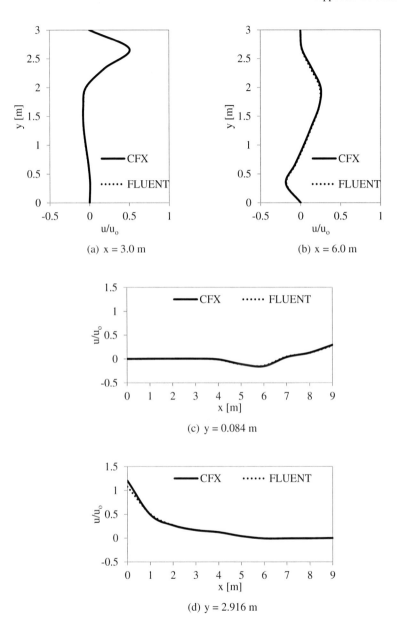

(a) x = 3.0 m

(b) x = 6.0 m

(c) y = 0.084 m

(d) y = 2.916 m

Figure B.4: Indoor air flow (laminar) with an emission source – Comparison of numerical simulation results (non-dimensional velocities) of the two CFD programs at the different lines locations (x = 3.0 m, x = 6.0 m, y = 0.084 m and y = 2.916 m)

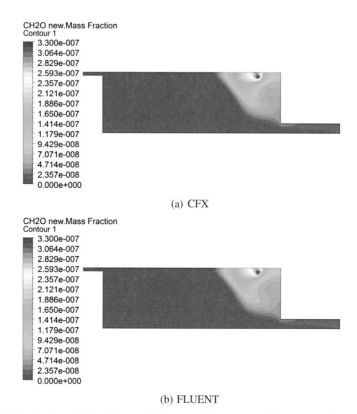

(a) CFX

(b) FLUENT

Figure B.5:   Indoor air flow (laminar) with an emission source – Comparison of mass
fraction distributions of formaldehyde inside the room from the two CFD
programs

As demonstrated in Figure B.5, the concentration of formaldehyde inside the room is relatively higher near the source and close to zero at the region near to the inlet where fresh air is
entering into the room. It can be concluded that air flow has a significant influence on the
distribution of formaldehyde emission inside the room.

B.1.3 OpenFOAM running in parallel on computing cluster

The indoor air flow model (turbulent case) presented in Section 4.1 was also simulated with OpenFOAM running in parallel on the computing cluster. The simulations were run with two different number of mesh nodes M1 = 361505 and M2 = 769230. OpenFOAM uses domain decomposition as the method of parallel computing. In the present work, OpenFOAM was run in parallel on the computing cluster of 12 nodes with 20 processors per node with a processor Intel® Xeon® CPU E5-2670v2 @2.50 GHz. The *simple* method of decomposition was employed in which the domain was split into pieces by direction.

The procedure adopted in this work to run OpenFOAM in parallel on the computing cluster is demonstrated in Figure B.6. The utility *decomposePar* was used to decompose the mesh and fields. A set of subdirectories were created for each processor in the case directory. The decomposed case was run in parallel using the MPI from Intel Composer SE Suite. A cluster script file was made which contains the information about the host names of the machines, the type of solver used, the number of nodes and processors etc. The utility *reconstructPar* was used to reconstruct the mesh and field data for the purpose of recreation of the complete domain and fields for the normal post-processing. All the boundary conditions, simulation parameters and other settings are the same as demonstrated in Table 4.1 of Section 4.1.1.

The results of the parallel performance analysis of OpenFOAM with one processor as a reference are shown in Figure B.7. The execution (computation) time generally decreases as the number of processors, on which the simulations are run, is increased as depicted in Figure B.7 a. The execution time decreases faster up to 5 numbers of processors per node and then decreases relatively slower with the further increase in the number of processors. The execution time is higher for the higher number of mesh nodes case (M2) as compared to that of lower number of mesh nodes case (M1). Figure B.7 b shows that the speedup generally increases with the increase in the number of processors per node. However, the speedup of M2 case is lower than that of M1 case. The parallel efficiency of OpenFOAM generally decreases with the increase in the number of processors per node as demonstrated in Figure B.7 c. The parallel efficiency for case M2 is always less than that of case M1.

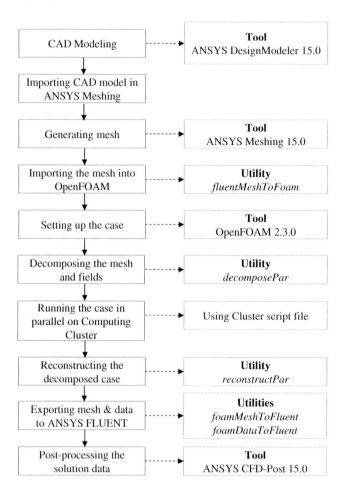

Figure B.6:   Procedure adopted to run OpenFOAM in parallel on the computing cluster

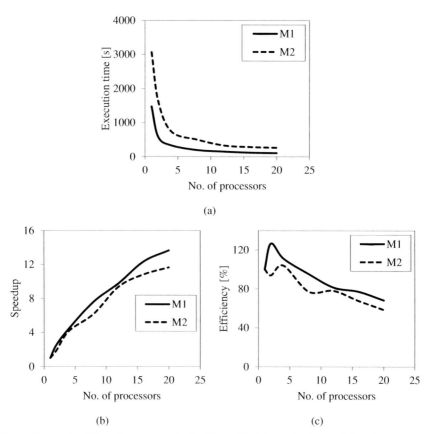

(a)

(b)

(c)

Figure B.7:   Parallel performance analysis of OpenFOAM running in parallel on the computing cluster

B.2   Developed simulation model based on double configuration

B.2.1   Time dependent expression for the mass source contribution

An example for the selection of the time dependent expression to specify the flow rate of the propane (mass source contribution) for case 1 when the measurement points are located near the distributor ($z = 3.375$ m) is explained in this section. The different trials of simulations were run with the different time dependent expressions of the boundary mass source contribution for propane as exhibited in Table B.3. The expression for which the average percentage difference between the simulated and the experimental concentrations at all the measurement points is relatively minimum is selected.

Table B.3:   Different trials of simulations for the selection of the best possible expression of the propane boundary mass source contribution

Simulation trial	Conditional statement	Time dependent expression for propane boundary source contribution
sim 1	if(t<1800[s], **expression**, 0[kg s^{-1}])	-(0.000000090[kg s^{-2}]*t)+0.000154[kg s^{-1}]
sim 2	if(t<1800[s], **expression**, 0[kg s^{-1}])	-(0.000000093[kg s^{-2}]*t)+0.000177[kg s^{-1}]
sim 3	if(t<1800[s], **expression**, 0[kg s^{-1}])	-(0.000000094[kg s^{-2}]*t)+0.000204[kg s^{-1}]
sim 4	if(t<1800[s], **expression**, 0[kg s^{-1}])	-(0.000000091[kg s^{-2}]*t)+0.000231[kg s^{-1}]
sim 5	if(t<1800[s], **expression**, 0[kg s^{-1}])	-(0.000000089[kg s^{-2}]*t)+0.000262[kg s^{-1}]
sim 6	if(t<1800[s], **expression**, 0[kg s^{-1}])	-(0.000000092[kg s^{-2}]*t)+0.000284[kg s^{-1}]

As shown in Figure B.8, the simulation results from sim 3 are in better agreement against the experimental data as compared to other simulation trial runs. For sim 3, the average percentage difference between simulated and experimental concentrations at point 1 and point 3 for the period when the propane is entering into the room is 16 % and 9.6 % respectively. Whereas, the average percentage difference between simulated and experimental concentrations for all the other simulation trial runs is greater than 16 % (at point 1) and 9.6 % (point 3). Hence, the best expression for propane boundary source is selected based on sim 3. The same methodology was adopted to determine the best expression for the specification of the component/s boundary source/s for all the other cases of two components simulation and three components

simulation. The determined best expressions are in accordance with the experimentally determined flowrate through the weight analysis as discussed in Section 5.3 of Chapter 5.

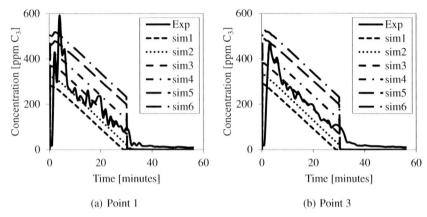

(a) Point 1 (b) Point 3

Figure B.8:   An example of the selection of the best expression for the specification of the
mass source contribution for propane

B.2.2   Influence of diffusion

This section demonstrates the influence of diffusion on the concentration distribution of the
pollutant in the room when the pollutant is entering into the room with the turbulent air flow.
The different trials of simulations are performed with the different values of the kinematic
diffusivity of the pollutant. The simulation case is the same Case 1 of the Two Components
System presented in Section 6.2.1 and the same simulation parameters, settings and boundary
conditions are used here. Table B.4 shows the different values of kinematic diffusivities
against the different simulation trial runs descriptions. As can be seen in Figure B.9, the concentration distributions of the pollutant at the two points inside the room are the same irrespective of the different kinematic diffusivities. The influence of diffusion on the concentration distribution of the pollutant in the room is negligible. The kinematic diffusivity of the
pollutant has little effect. Hence, it can be concluded that the present problem of the spread,
transportation and distribution of the pollutant with the turbulent air flow is considered to be a
convection dominated phenomena. The convection currents transport the pollutant along the
streamlines.

Table B.4:   Different values of kinematic diffusivities against the different simulation trial descriptions

Simulation description	Diffusivity ($m^2 s^{-1}$)
Sim_1	0.000005
Sim_2	0.000012
Sim_3	0.000020
Sim_4	0.000025
Sim_5	0.000030
Sim_6	0.000040
Sim_7	0.000060

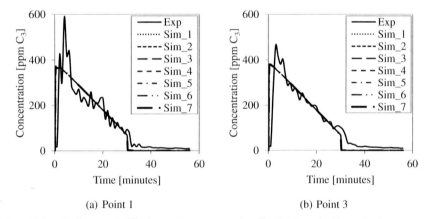

(a) Point 1   (b) Point 3

Figure B.9:   Influence of diffusion on the concentration distribution of pollutant in the room

B.2.3   Comparison of simulation results with experimental results

Two components system (without flow obstruction)

Case – Measurement points away from the distributor (z = 1.125 m)

The experimental results along with the measurement locations for this case when the measurement points are located away from the distributor have already been presented in Appendix A.3.1. The simulation boundary conditions, parameters and other settings have already been mentioned in Section 6.2.1. The comparison of the measured and simulated concentration profiles of propane inside the room at the three measurement points located away from the distributor exhibited in Figure B.10 for a period of 2880 s demonstrates that the simulation and experimental results generally show the same response with time. Figure B.10 a shows that the difference between simulation and experimental concentrations at point 1 at 16 minutes and 22 minutes are 0.2 ppm and 12 ppm respectively. The average percentage difference between simulated and experimental concentrations at point 1 between 3 minutes and 22 minutes when the propane is entering into the room is 11.05 %.

Figure B.10 b demonstrates that the difference between simulation and experimental concentrations at point 2 at 16 minutes and 22 minutes are 18.4 ppm and 32 ppm respectively. Whereas, the average percentage difference between simulated and experimental concentrations at point 2 between 3 minutes and 22 minutes is 10.22 % for the period when the propane is entering into the room. The difference between simulation and experimental concentrations at point 3 at 16 minutes and 22 minutes are 17.8 ppm and 46.2 ppm respectively as shown in Figure B.10 c. The average percentage difference between simulated and experimental concentrations at point 3 between 3 minutes and 22 minutes is 10.76 % for the period when the propane is entering into the room.

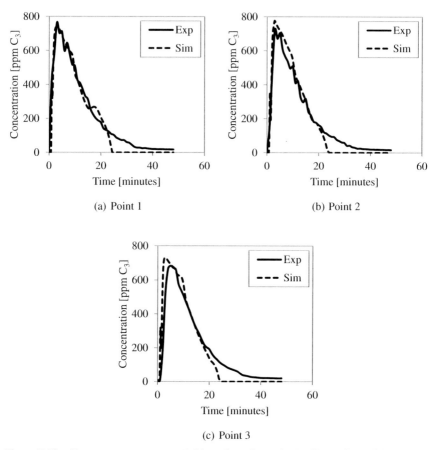

(a) Point 1

(b) Point 2

(c) Point 3

Figure B.10: Two components system (without flow obstruction) – Comparison of the simulation results with the measurement results for the case when the measurement points are located away from the distributor (z = 1.125 m)

Two components system (with flow obstruction)

Case – Relative Distance

In this section, the simulation results of the two components system are compared with the corresponding experimental results for the purpose of validation of the developed simulation model when the flow obstruction is present in the room for the case 1 "Relative Distance" of Appendix A.3.2 . The developed simulation model based on the double configuration already explained in Chapter 3, implemented in Section 4.3 of Chapter 4 and validated in Chapter 6 is employed to compare the simulation results with the experimental results for the two components system when the flow obstruction is present in the room. The inlet mass flow rate of air for this case is 0.31 kg s^{-1}. Table B.5 shows the mass source contribution for propane for this case specified by using the same methodology as demonstrated in Section 6.2.1 of Chapter 6. The total time duration of the simulation for this case is 105 minutes. All the other boundary conditions, simulation parameters and settings are the same as mentioned in Section 6.2.1 of Chapter 6. The comparison of the propane concentration profiles exhibited in Figure B.11 a at the measurement point 2 shows that the difference between simulation and experimental concentrations at 19 minutes and 40 minutes are 25.25 ppm and 8.84 ppm respectively. The average percentage difference between simulated and experimental concentrations at point 2 between 3 minutes and 30 minutes when the propane is entering into the room is 12.27 %.

Table B.5:   Two components system (with flow obstruction) – Boundary mass source contribution of propane for the case "Relative Distance"

Conditional statement	expression
if(t<1800[s], **expression**, 0[kg s^{-1}])	-(0.00000015[kg s^{-2}]*t)+0.00061[kg s^{-1}]

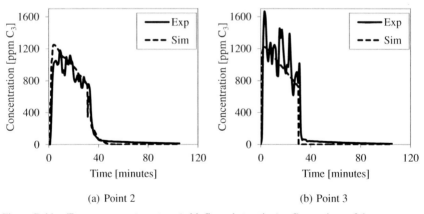

(a) Point 2   (b) Point 3

Figure B.11:   Two components system (with flow obstruction) – Comparison of the simulation results with the measurement results for the case "Relative Distance"

The difference between simulation and experimental concentrations at point 3 at 19 minutes and 40 minutes are 55.95 ppm and 41.29 ppm respectively as demonstrated in Figure B.11 b. The simulation and experimental results generally show the same response with time. For the period when the propane is entering into the room, the average percentage difference between simulated and experimental concentrations at point 3 between 3 minutes and 30 minutes is 15.47 %.

References

[1] "India Environment Portal," [Online]. Available: http://www.indiaenvironmentportal.org.in/category/2543/thesaurus/gas-leakages. [Accessed 16 September 2015].

[2] [Online]. Available: http://news.xinhuanet.com/english/2015-08/28/c_134566134.htm. [Accessed 16 September 2015].

[3] [Online]. Available: http://www.mirror.co.uk/news/world-news/see-horrific-650ft-towering-inferno-4493746. [Accessed 17 September 2016].

[4] [Online]. Available: http://www.thehindu.com/news/national/other-states/ammonia-gas-tanker-leak-in-ludhiana-punjab/article7312213.ece. [Accessed 15 September 2015].

[5] [Online]. Available: http://www.thehindu.com/news/national/telangana/five-injured-in-fire-accident-after-gas-cylinder-leak/article7108349.ece. [Accessed 15 September 2015].

[6] [Online]. Available: http://www.dawn.com/news/1167074. [Accessed 15 September 2015].

[7] [Online]. Available: http://www.dailymail.co.uk/news/article-2967209/Natural-gas-blast-destroys-home-injures-5-New-Jersey.html. [Accessed 15 September 2015].

[8] [Online]. Available: http://articles.economictimes.indiatimes.com/2015-02-22/news/59386556_1_toxic-gas-leak-power-generator-school-employee. [Accessed 15 September 2015].

[9] [Online]. Available: http://www.firehouse.com/news/12033474/propane-leak-sparks-deadly-victoria-texas-house-explosion-fire. [Accessed 15 September 2015].

151

[10] [Online]. Available: http://arynews.tv/en/three-killed-gas-leakage-incidents/. [Accessed 15 September 2015].

[11] [Online]. Available: http://www.dailyexcelsior.com/three-injured-fire-caused-due-lpg-leak/. [Accessed 15 September 2015].

[12] [Online]. Available: http://www.thehindu.com/news/cities/bangalore/man-killed-wife-injured-in-lpg-leakage/article5868295.ece. [Accessed 15 September 2015].

[13] [Online]. Available: http://dunyanews.tv/index.php/en/Pakistan/200488-Rawalpindi-Five-injured-in-gas-leakage-blast. [Accessed 15 September 2015].

[14] [Online]. Available: http://www.thenews.com.pk/article-90227-Lahore-Nine-labours-faint-after-gas-leakage. [Accessed 15 September 2015].

[15] S. Kirchner, F. Maupetit, D. Quenard, P. Rouxel and D. Giraud, "Characterization of adsorption/desorption of volatile organic compounds in indoor surface materials," in *Proceedings of Healthy Building*, Milano, Italy, 1995.

[16] T. Yoshida, I. Matsunaga, K. Tomioka and S. Kumagai, "Interior air pollution in automotive cabins by volatile organic compounds diffusing from interior materials: I. Survey of 101 types of Japanese domestically produced cars for private use," *Indoor Built Environment,* vol. 15, pp. 425-444, 2006.

[17] Air Quality Sciences, Inc., "Indoor air quality hazards of new cars," 2006.

[18] H. Nourbakhsh, D. Mowla and F. Esmaeilzadeh, "Predicting the three dimensional distribution of gas pollutants for industrial-type geometries in the South Pars Gas Complex using Computational Fluid Dynamics," *Ind. Eng. Chem. Res.,* vol. 52, p. 6559–6570, 2013.

[19] C. Spengler, "Experimentelle und numerische Ermittlung von Verdunstungsemissionen im Luftansaugsystem von Motoren," Logos Verlag Berlin GmbH, Germany, 2012.

[20] C. Spengler, C. Merten and F. Pfeiffer, "Measuring method and simulation model to determine the spatial and temporal distribution of evaporative emissions in the air intake system of Otto-engines," in *Proceedings of 7th European Congress of Chemical*

Engineering & 19th International Congress of Chemical and Process Engineering CHISA, Prague, Tschechien, 28 Aug - 1 Sept 2010.

[21] C. Spengler and C. Merten, "Messverfahren und Simulationsmodell zur Ermittlung der zeitlichen und örtlichen Verteilung der Verdunstungsemissionen im Luftansaugsystem von Ottomotoren," in *Proceedings of ANSYS Conference & 28th CADFEM Users' Meeting*, Aachen, Deutschland, 3-5 Nov 2010.

[22] C. Spengler and C. Merten, "Simulation of the temporal and spatial distribution of evaporative emissions inside the air-intake system of Otto-engines," Technischer Bericht, Institut für Chemische Verfahrenstechnik, Universität Stuttgart, Deutschland, 2011.

[23] J. D. Posner, C. R. Buchanan and D. D. Rankin, "Measurement and prediction of indoor air flow in a model room," *Energy and Buildings,* vol. 35, pp. 515-526, 2003.

[24] F. Kuznik, G. Rusaouen and J. Brau, "Experimental and numerical study of a full scale ventilated enclosure: Comparison of four two equations closure turbulence models," *Building and Environment,* vol. 42, p. 1043–1053, 2007.

[25] Z. J. Zhai, Z. Zhang, W. Zhang and Q. Y. Chen, "Evaluation of Various Turbulence Models in Predicting Airflow and Turbulence in Enclosed Environments by CFD: Part 1—Summary of Prevalent Turbulence Models," *HVAC&R Research,* vol. 13, pp. 853-870, 2007.

[26] Z. Zhang, W. Zhang, Z. J. Zhai and Q. Y. Chen, "Evaluation of Various Turbulence Models in Predicting Airflow and Turbulence in Enclosed Environments by CFD: Part 2—Comparison with Experimental Data from Literature," *HVAC&R Research,* vol. 13, pp. 871-886, 2007.

[27] Y. Wu, E. Yu and Y. Xu, "Simulation and analysis of indoor gas leakage," in *Proceedings of IBPSA Conference*, Beijing, China, 3-6 September 2007.

[28] F. Gavelli, S. G. Davis and O. R. Hansen, "A modern tool for the investigation of indoor flammable gas migration," in *ISFI meeting*, Hyattsville, USA, September 27-29 2010.

[29] C. D. Barley, K. Gawlik, J. Ohi and R. Hewett, "Analysis of buoyancy-driven ventilation of hydrogen from buildings," in *2nd International Conference on Hydrogen Safety*, San

Sebastian, Spain, September 11-13, 2007.

[30] M. Siddiqui, S. Jayanti and T. Swaminathan, "CFD analysis of dense gas dispersion in indoor environment for risk assessment and risk mitigation," *Journal of Hazardous Materials,* Vols. 209-210, pp. 177-185, 2012.

[31] Y. Muharam and H. Septian, "Simulation of gas leakage in a gas utilization system in household sector," *International Journal of Technology,* vol. 3, pp. 224-231, 2013.

[32] R. S. Nagaosa, "A new numerical formulation of gas leakage and spread into a residential space in terms of hazard analysis," *Journal of Hazardous Materials,* vol. 271, pp. 266-274, 2014.

[33] K. C. Chung and S. P. Hsu, "Effect of ventilation pattern on room air and contaminant distribution," *Building and Environment,* vol. 36, p. 989–998, 2001.

[34] K. W. D. Cheong, E. Djunaedy, T. K. Poh, K. W. Tham, S. C. Sekhar, N. H. Wong and M. B. Ullah, "Measurements and computations of contaminant's distribution in an office environment," *Building and Environment,* vol. 38, p. 135 – 145, 2003.

[35] Z. Zhang, X. Chen, S. Mazumdar, T. Zhang and Q. Chen, "Experimental and numerical investigation of airflow and contaminant transport in an airliner cabin mockup," *Building and Environment,* vol. 44, p. 85–94, 2009.

[36] B. Gu, O. Kaschtschejewa and M. Schmidt, "D6.3 Comparison of simulation results with measured results," CETIEB - GA No. 285623, Germany, 2014.

[37] T. Schlosser, O. Kaschtschejewa and B. Gu, "D6.1 Information about the air flow pattern and the distribution of air pollutants," CETIEB GA No. 285623, Germany, 2014.

[38] K. C. Chung, "Three-dimensional analysis of airflow and contaminant particle transport in a partitioned enclosure," *Building and Environment,* vol. 34, pp. 7-17, 1999.

[39] Z. F. Tian, J. Y. Tu, G. H. Yeoh and R. K. K. Yuen, "On the numerical study of contaminant particle concentration in indoor airflow," *Building and Environment,* vol. 41, pp. 1504-1514, 2006.

[40] Z. Zhang and Q. Chen, "Experimental measurements and numerical simulations of particle transport and distribution in ventilated rooms," *Atmospheric Environment,* vol. 40, pp. 3396-3408, 2006.

[41] Z. F. Tian, J. Y. Tu, G. H. Yeoh and R. K. K. Yuen, "Numerical studies of indoor airflow and particle dispersion by large Eddy simulation," *Building and Environment,* vol. 42, pp. 3483-3492, 2007.

[42] S. Murakami, S. Kato and K. Ito, "Coupled analysis of VOCs emission and diffusion in a ventilated room by CFD," in *EPIC,* Lyon, France, 19-21 November 1998.

[43] S. Murakami, S. Kato, K. Ito, A. Yamamoto, Y. Kondo and J.-i. Fujimura, "Chemical Pollutants Distribution in a Room Based on CFD Simulation Coupled with Emission/Sorption Analysis," *ASHRAE Transaction,* 2001.

[44] S. Murakami, S. Kato, K. Ito and Q. Zhu, "Modeling and CFD prediction for diffusion and adsorption within room with various adsorption isotherms," *Indoor Air,* vol. 13, pp. 20-27, 2003.

[45] C. N. Kim, W. H. Choi, S. J. Choung, C.-H. Park and D. S. Kim, "Efficient ventilation of VOC spread in a small-scale painting process," *Building and Environment,* pp. 1321-1328, 2002.

[46] B. Deng and C. N. Kim, "A new CFD model for VOC emission based on the General Adsorption Isotherm," *JSME International Journal,* vol. 47, no. 2, pp. 396-402, 2004.

[47] B. Deng, C. N. Kim and F. Zhang, "Numerical simulation of VOCs distribution in a room with a new carpet," *Heat Mass Transfer,* vol. 43, pp. 975-983, 2007.

[48] B. Deng and C. N. Kim, "CFD simulation of VOCs concentrations in a resident building with new carpet under different ventilation strategies," *Building and Environment,* vol. 42, pp. 297-303, 2007.

[49] G. Iaccarino, "Predictions of a Turbulent Separated Flow Using Commercial CFD Codes," *Journal of Fluids Engineering,* vol. 123, pp. 819-828, 2001.

[50] P. Kopyt and W. Gwarek, "A Comparison of Commercial CFD Software Capable of

Coupling to External Electromagnetic Software for Modeling of Microwave Heating Process," Institute of Radioelectronics, Warsaw University of Technology, Warsaw, Poland, 2004.

[51] S. R. Hanna, M. J. Brown, F. E. Camelli, S. T. Chan, W. J. Coirier, O. R. Hansen, A. H. Huber, S. Kim and R. M. Reynolds, "Detailed simulations of atmospheric flow and dispersion in downtown Manhattan - An application of five computational fluid dynamics models," American Meteorological Society, 2006.

[52] A. Horvath, C. Jordan and M. Harasek, "Comparison of a Commercial and an Open Source CFD Software Package," Institute of Chemical Engineering, Vienna University of Technology, Austria, 2006.

[53] G. H. Lee, Y. S. Bang and S. W. Woo, "Performance Comparison of the Commercial CFD Software for the Prediction of Turbulent Flow through Tube Bundles," in *Transactions of the Korean Nuclear Society Autumn Meeting*, Gyeongju, Korea, 2012.

[54] M. R. Haghgoo, "A comparison of CFD Software packages to find the suitable one for numerical modeling of gasification process (Progress Report)," Department of Computer Science, University of Saskatchewan, Canada, 2013.

[55] A. Mackenzie, A. Lopez, K. Ritos, M. T. Stickland and W. M. Dempster, "A comparison of CFD software packages' ability to model a submerged jet," in *Eleventh International Conference on CFD in the Minerals and Process Industries*, CSIRO, Melbourne, Australia, 2015.

[56] T. Vogeltanz, "Comparison of open-source CFD software for aerodynamic analysis of mini-UAV," in *34th Digital Avionics Systems Conference (DASC)*, Prague, Czech Republic, 2015.

[57] N. Li, "Comparison between three different CFD software and numerical simulation of an ambulance hall (Master thesis)," KTH Industrial Engineering and Management, Stockholm, Austria, 2015.

[58] P. V. Nielsen, "Specification of a two-dimensional test case," Department of Building Technology and Structure Engineering. Aalborg University Denmark, 1990.

[59] ANSYS, Inc., "DesignModeler User's Guide," Canonsburg, PA USA, November 2013.

[60] ANSYS, Inc., "ANSYS CFX Introduction," Canonsburg, PA USA, November 2013.

[61] ANSYS, Inc., "ANSYS Fluent User's Guide," Canonsburg, PA, November 2013.

[62] ANSYS, Inc., "ANSYS Meshing User's Guide," Canonsburg, PA USA, November 2013.

[63] H. K. Versteeg and W. Malalasekera, An introduction to computational fluid dynamics - The finite volume method, England: Longman Scientific & Technical, 1995.

[64] ANSYS, Inc., "ANSYS CFX-Solver Theory Guide," Canonsburg, PA USA, November 2013.

[65] ANSYS, Inc., "ANSYS CFX-Solver Modeling Guide," Canonsburg, PA USA, November 2013.

[66] B. E. Launder and D. B. Spalding, "The numerical computation of turbulent flows," Computer Methods in Applied Mechanics and Engineering, vol. 3, p. 269–289, March 1974.

[67] N. Matsunaga, M. Hori and A. Nagashima, "Gaseous Diffusion Coefficients of Propane and Propylene into Air, Nitrogen and Oxygen," Netsu Bussei, vol. 21, no. 3, pp. 143-148, 2007.

[68] T. R. Marrero and E. A. Mason, "Gaseous diffusion coefficients," Physical and Chemical Reference Data, vol. 1, 1972.

[69] National Institute of Standards and Technology, "NIST Reference Fluid Thermodynamic and Transport Properties - REFPROP," Boulder, Colorado USA, April 2007.

[70] OpenFOAM Foundation, "OpenFOAM - User Guide," February 2014.

[71] ANSYS, Inc., "ANSYS CFD-Post User's Guide," Canonsburg, PA USA, November 2013.

[72] ANSYS, Inc., "Workbench User's Guide," Canonsburg, PA USA, November 2013.

References

[73] ANSYS, Inc., "ANSYS CFX-Pre User's Guide," Canonsburg, PA, November 2013.

[74] ANSYS, Inc., "ANSYS CFX-Solver Manager User's Guide," Canonsburg, PA USA, November 2013.

[75] ANSYS, Inc., "ANSYS Fluent Getting Started Guide," Canonsburg, PA USA, November 2013.

[76] N. Zheng, "Simulation des Konzentrationsverlaufs von emittierendem Formaldehyd in einer Emissionsprüfkammer (Masters Thesis)," Institute of Chemical Process Engineering, University of Stuttgart, Germany, 2014.

[77] SK-Elektronik GmbH, [Online]. Available: http://sk-elektronik.biz. [Accessed 15 Ocotber 2015].

[78] SK-Elektronik GmbH, "Data Sheet: Thermo-FID TG".

[79] ABB Automation GmbH, [Online]. Available: http://new.abb.com/de. [Accessed 15 October 2015].

[80] ABB, "EasyLine EL3000 Series Data Sheet," 2006.

[81] Halstrup-Walcher, [Online]. Available: https://www.halstrup-walcher.de/en/index.php. [Accessed 10 October 2015].

[82] Halstrup-Walcher, "Instruction Manual for P26 Differential Pressure Transducer," Germany, 2015.

[83] L. Rong and V. P. Nielsen, "Simulation with different turbulence models in an annex 20 room benchmark test using Ansys CFX 11.0 (Technical Report No. 46)," Aalborg University, Denmark, 2008.

[84] L. J. Dréau, P. Heiselberg and V. P. Nielsen, "Simulation with different turbulence models in an Annex 20 benchmark test using Star-CCM+ (Technical Report No. 147)," Aalborg University, Denmark, 2012.

[85] [Online]. Available: http://www.engineeringtoolbox.com/explosive-concentration-limits-d_423.html. [Accessed 18 December 2015].

[86] [Online]. Available: http://www.engineeringtoolbox.com/air-change-rate-room-d_867.html. [Accessed 18 December 2015].

[87] [Online]. Available: https://www.osha.gov/dts/chemicalsampling/data/CH_264000.html. [Accessed 18 December 2015].

[88] [Online]. Available: http://www.k2bw.com/5_c_18.htm. [Accessed 20 December 2015].

[89] ABB, "Special Report - Instrumentation & Analytics," 2006.

[90] Halstrup Walcher, "In air ducts and processes - Measurement of volume flows," 2015.